Principle, Application, and Advancement of Temperature and Humidity Sensors

Edited by Mingqiang Yi and Fenglei Niu

Published in London, United Kingdom

Principle, Application, and Advancement of Temperature and Humidity Sensors
http://dx.doi.org/10.5772/intechopen.1009701
Edited by Mingqiang Yi and Fenglei Niu

Contributors
A.M. Orheruata, Alexandre Bessa dos Santos, Alexandre Cotta, Enzo Dantas da Silva, Fatih Evren, Fenglei Niu, Flávio Augusto Marques, Haicai Lyu, Jefferson E. Tsuchida, Jonas Henrique Osório, João Gabriel de Carvalho, Juan C. Rubio, Lifan Yang, Luiz David Silva, Martin Bolduc, Mingqiang Yi, O.D. Asemota, O.E. Okhale, Pengcheng Yang, Renato Faraco Filho, Sinei Gonçalves Júnior, Xianjun Chen, Xiaoyang Lun

First published in London, United Kingdom, 2026 by IntechOpen
IntechOpen is the global imprint of INTECHOPEN LIMITED, registered in England and Wales, registration number: 11086078, 167-169 Great Portland Street, London, W1W 5PF, United Kingdom

For EU product safety concerns: IN TECH d.o.o., Prolaz Marije Krucifikse Kozulić 3, 51000 Rijeka, Croatia, info@intechopen.com or visit our website at intechopen.com.

British Library Cataloguing-in-Publication Data
A catalogue record for this book is available from the British Library

Principle, Application, and Advancement of Temperature and Humidity Sensors
Edited by Mingqiang Yi and Fenglei Niu
p. cm.
Print ISBN 978-1-83635-944-9
Online ISBN 978-1-83635-943-2
eBook (PDF) ISBN 978-1-83635-945-6

Meet the editors

Dr. Mingqiang Yi has served as the CEO of Microfluidic Foundry since 2011. He has a broad background in micro- and nanotechnology, as well as sensor development. From 2009 to 2011, Dr. Yi served as the process engineering manager for Alphabet Energy, Inc., where he was responsible for developing silicon nanowire-based thermoelectric modules. From 2005 to 2008, he served as the project leader for RheoSense, Inc., where he was responsible for developing microfluidic chip-based viscometers. From 2001 to 2002, Dr. Yi was a MEMS scientist at Los Gatos Research, Inc., responsible for the development of high-pressure electrokinetic micropumps. From 2002 to 2005, he was a postdoctoral fellow at the University of California, Berkeley, USA, where he developed microfluidic sensors for detecting DNA hybridization and enzymatic reaction products as well as for protein analysis. Dr. Yi obtained his Ph.D. in Mechanical Engineering from the University of Pennsylvania, USA, in 2001, and a BEng and MEng from Tsinghua University in 1994 and 1997, respectively.

Dr. Fenglei Niu is a pioneering researcher in nuclear thermal hydraulics. He is a former dean and professor of the School of Nuclear Science and Engineering, North China Electric Power University, and a member of the American Nuclear Society. Dr. Niu has multiple registered patents to his credit. He obtained his Ph.D. in Nuclear Engineering from the University of California, Berkeley, in 2004, and his BEng and MEng from Tsinghua University in 1994 and 1997, respectively.

Contents

Preface

Temperature and humidity sensors are fundamental components in a wide range of modern technologies, playing critical roles in industries such as environmental monitoring, industrial automation, healthcare, and consumer electronics. The measurement and control of temperature and humidity are crucial aspects of modern life. From industrial automation to nuclear energy systems, from agricultural monitoring to biomedical applications, these sensors play a vital role in enhancing performance, safety, and sustainability. As the world becomes increasingly interconnected and we transition from the "Age of Information" to the "Age of Intelligence," driven by AI, the demand for accurate and reliable temperature and humidity sensors continues to grow. This book offers a comprehensive examination of the latest advancements in these essential sensing devices, encompassing their underlying principles, diverse technological implementations, broad-ranging applications, and their future prospects.

This book aims to inspire innovation and highlight areas where research and development are actively pushing boundaries. This book offers an in-depth examination of the latest advancements in temperature and humidity sensor technology, including the development of new materials, novel fabrication techniques, and their expanding applications. We also examine the challenges and opportunities facing the field, including the need for improved accuracy, reliability, and energy efficiency. The application of AI (artificial intelligence) in temperature and humidity sensors primarily occurs in the fields of smart homes, industrial automation, and medical equipment. In industrial automation, temperature and humidity sensors can be combined with AI technology for high-precision measurement and control. In medical equipment, temperature and humidity sensors can be combined with AI to enable high-precision measurements and real-time monitoring.

In Chapter 1, "Printed Flexible Sensors for Low-Cost Temperature and Humidity Monitoring: Materials, Processes, and Applications", Juan Rubio and Martin Bolduc present a comprehensive and logically structured overview of printed flexible sensors for low-cost temperature and humidity monitoring, discussing their materials, fabrication processes, and typical applications. Printed flexible temperature and humidity sensors are an emerging technology, representing a distinct approach from their semiconductor counterparts. The printed sensors enable numerous novel applications for temperature and humidity sensing, such as wearable devices and integration into complex shapes.

In Chapter 2, "Fiber Optics Technology for Temperature and Humidity Sensor Development", Renato Faraco Filho, João Gabriel de Carvalho, Luiz David Silva, Enzo da Silva, Sinei Gonçalves Junior, Flávio Augusto Marques, Alexandre Cotta, Alexandre Bessa dos Santos, Jonas Henrique Osório and Jefferson Tsuchida provide an overview of fiber optic sensors for temperature and humidity monitoring, detailing their fundamental principles and advanced architectures. Five different sensing platforms are investigated, namely fiber Bragg gratings, long-period gratings, Fabry-Perot interferometers, multimode interferometry sensors, and microstructured optical fibers. This review emphasizes the versatility of fiber optic sensors and their significant advantages in sensing applications, including immunity to electromagnetic interference, compactness, and the ability to operate in harsh environments. Continued developments in microstructure modification, liquid penetration techniques, and novel functional coatings are opening up new possibilities for creating compact, highly sensitive, and fast-response sensing platforms that can support a variety of emerging sensing technologies.

In Chapter 3, "Temperature and Humidity Sensors' Application in Power Equipment", Lifan Yang describes the application of sensors on specific power equipment, such as switchgear, transformers, and cables. The chapter introduces the various applications of temperature and humidity sensors in power equipment and examines the service life of these sensors. Many technologies have been applied to temperature measurement in various conditions, including passive wireless temperature measurement (e.g., surface acoustic wave-based sensing), passive wired temperature measurement (e.g., optical fiber sensing), active wireless temperature measurement (e.g., inductively powered wireless sensing), and infrared temperature measurement.

In Chapter 4, "Seasonal Variations in Temperature-Humidity Index (THI) and Their Effects on Physiological, Reproductive, and Productivity Performance of Rabbit Does Raised in Benin City, Nigeria", Dennis Osarenren Asemota, O.E. Okhale and Arierhire Michael Orheruata studied the effects of the Temperature-Humidity Index (THI) on rabbit does in Benin City, Nigeria. The findings highlight the physiological and reproductive challenges rabbits face in this climate, necessitating the implementation of mitigation strategies for sustainable rabbit farming. Temperature and humidity sensors are critical for data collection to study the impact of heat stress on rabbit productivity in a humid tropical environment. The accurate sensing of temperature and humidity is critically important for studying the effects of THI on rabbit does. The THI is a calculated value derived directly from these two environmental parameters. Accurate and consistent temperature and humidity sensing is the fundamental backbone of the study. It transforms anecdotal observations of heat stress into quantifiable, scientific data, allowing for meaningful analysis and the development of effective mitigation strategies for rabbit farming in tropical climates.

In Chapter 5, "Contactless Temperature Sensors for Enhancing Thermal Comfort Measurements in Buildings", Fatih Evren provides a timely and comprehensive overview of contactless temperature sensing technologies and their applications in thermal comfort measurements within buildings. This chapter addresses both fundamental theory and recent technological developments, making it suitable for a broad audience in building science, HVAC engineering, and smart building systems. Contactless sensors will be a crucial component of smart building systems, enabling real-time monitoring and control. Contactless temperature sensing offers a promising route to enhance the precision, responsiveness, and occupant-centricity of thermal comfort measurement in buildings. Contactless temperature sensing addresses many limitations of conventional contact or ambient measurements, particularly for skin temperature, spatial variation, and user comfort.

In Chapter 6, "Recent Application of Advanced Temperature Sensors in Special Scenarios of Nuclear Energy Research", Mingqiang Yi, Fenglei Niu, Haicai Lyu, Xiaoyang Lun, Xianjun Chen and Pengcheng Yang address an important and timely topic in nuclear energy research: advanced methodologies for temperature measurement in environments characterized by high temperature, pressure, and irradiation. Accurate thermal monitoring remains critical for both operational safety and efficiency in current and next-generation reactor systems. This review appropriately considers four major classes of sensors (engineered sheathed thermocouples, fiber-optic sensors, passive thermoacoustic sensors, and thermochromic liquid crystal thermography) and situates each within relevant application scenarios, including liquid-metal and gas-cooled reactor environments.

Through in-depth discussions of real-world applications, this book illustrates the crucial role of temperature and humidity sensors in addressing critical challenges in environmental control, industrial processes, and human health and well-being. This resource serves as an invaluable guide for researchers, engineers, and practitioners seeking to advance their knowledge and expertise in this crucial field. Whether you are a researcher, engineer, or student, this book aims to provide a valuable resource for understanding the principles, applications, and advancements of temperature and humidity sensors. By exploring the latest developments and trends in this field, we hope to inspire innovation and advancement in the years to come. We hope that this book serves as both a reference and a source of inspiration for all those who seek to better measure and manage the world around us.

The editors extend their heartfelt thanks to the distinguished authors for their invaluable contributions to this book. Their chapters added both depth and clarity to the overall narrative, and their expertise brought a unique perspective that significantly enhanced the work. It has been a pleasure working with all the authors through the editorial process. We thank them for their professionalism, responsiveness, and thoughtful revisions. Their efforts have helped shape a book that we can all be proud of.

The editors would also like to extend their sincere thanks to Publishing Process Manager Ms. Papuga for her outstanding support throughout the development and production of this book. Her coordination, communication, and attention to detail made a meaningful difference at every stage of the process. From managing timelines and liaising with contributors to overseeing the many behind-the-scenes tasks that keep everything moving smoothly, her efforts ensured that the project stayed on track and that our authors felt supported and informed.

The editors would like to thank their family members for their unwavering support throughout the editing process of this book. Their patience, encouragement, and understanding made it possible for us to dedicate the long hours and energy required to bring this project to fruition. The academic editor, Mingqiang Yi, is deeply appreciative of the steady presence and the great strength of his wife, Mrs. Yunju Bao, his son, Aili Yi, and his daughter, Aimi Yi. The co-editor, Fenglei Niu, is truly thankful for his wife's unwavering belief and encouragement. This book may have our name on the cover, but their support is woven into every chapter. We are endlessly grateful.

Mingqiang Yi and Fenglei Niu
Microfluidic Foundry,
San Pablo, California, USA

Chapter 1

Printed Flexible Sensors for Low-Cost Temperature and Humidity Monitoring: Materials, Processes, and Applications

Juan C. Rubio and Martin Bolduc

Abstract

Printed flexible temperature and humidity sensors have emerged as essential components for diverse applications requiring low-cost, scalable, and conformable environmental monitoring. This chapter presents a comprehensive overview of recent advancements in printed flexible sensors for temperature and humidity, emphasizing material selection, additive fabrication techniques, and practical applications. The chapter analyzes critical printing processes such as screen printing, inkjet printing, aerosol jet printing, gravure, and roll-to-roll methods, assessing their impact on feature resolution, ink properties, and processing temperatures. Comparative analyses highlight performance trends across metal nanoparticle inks, conductive polymers, nanocomposites, and two-dimensional materials, clearly linking these materials' stability, response time, and mechanical durability to specific fabrication conditions and substrate characteristics. Additionally, the chapter explores strategies for effective encapsulation, calibration, and artificial intelligence (AI)-supported drift correction to enhance sensor accuracy and reliability. By summarizing practical design guidelines and market outlooks, the chapter provides clear pathways for scaling laboratory prototypes into fully deployable sensor system configurations. The insights presented demonstrate the significant potential of printed flexible sensors in healthcare wearables, environmental monitoring networks, and industrial internet of things (IoT) solutions, contributing to transformative advancements in sensing technology.

Keywords: printed flexible sensors, temperature sensing, humidity sensing, screen printing, inkjet printing, nanomaterials, additive manufacturing

1. Introduction

Monitoring temperature and humidity is essential across diverse modern applications. In healthcare and biomedical monitoring, flexible temperature and humidity sensors enable continuous patient vital-sign tracking and advanced diagnostics. Within industrial automation, intelligent sensors support real-time process control, predictive maintenance, and safety assurance in smart factories [1]. For food storage and cold-chain logistics, internet of things (IoT)-based sensors maintain optimal conditions to reduce spoilage and ensure product quality [2]. In weather forecasting and building climate control, optical and electronic humidity sensors provide accurate environmental data for HVAC systems and meteorological stations [3]. Finally, in transportation and IoT deployments, low-power humidity and temperature sensors facilitate condition monitoring in vehicles, cargo holds, and smart-city infrastructures [4].

Printed flexible sensor technology has emerged as a promising approach to meet the growing demand for low-cost, scalable, and ready-to-deploy sensors in these areas. Printed sensors are manufactured via additive techniques (for example, screen printing or inkjet printing of electronic inks) that pattern functional materials directly onto substrates [1]. This approach radically simplifies fabrication compared to traditional silicon microfabrication, requiring fewer process steps and no complex etching stages, avoiding material waste and hazardous chemicals.

The key advantages of printed electronics for sensor development include dramatically reduced production costs, on the order of one-tenth to one-hundredth of the investment required for conventional silicon technology, and shorter production times, as well as increased manufacturing area and throughput [5, 6]. Additionally, printed sensors can be made on flexible substrates, granting mechanical durability with bending/stretching endurance and allowing electronics to conform to curved or soft surfaces [7, 8]. In practical terms, flexible printed sensors can be embedded in clothing, wearables, or structural materials, enabling applications and form factors not achievable with rigid sensors. Furthermore, unlike many traditional sensors that require expensive, specialized equipment, printed sensors can be produced with relatively simple and cost-effective processes, offering excellent production efficiency and adaptability to various surfaces [9].

According to the report by IDTechEx, the market for fully printed sensors is projected to reach approximately USD 4.5 billion by 2030. Humidity and temperature sensors are highlighted as the fastest-growing segment, driven by the development of novel composite inks incorporating carbon nanotubes and silicon nanoparticles. These advancements enable the fabrication of high-resolution, conformable sensor arrays ideal for applications such as IoT-based environmental monitoring, real-time skin-condition tracking, and battery temperature control. These printed devices offer considerable potential by combining low-cost production, large-area deposition, and ease of integration into flexible and wearable systems [10].

This chapter provides an overview of printed flexible sensors for low-cost temperature and humidity monitoring, discussing materials, fabrication processes, and representative applications. The focus is primarily on printed resistive sensors, such as thermistor-type temperature sensors and resistive humidity sensors, manufactured using specialized functional materials and conductive inks. Common sensing materials are presented, including silver nanoparticle-based inks for electrodes, conductive polymers employed as temperature-sensitive resistive films, printable nanocomposites designed for humidity absorption, and discuss their integration into flexible substrates such as polyimide (Kapton®), polyethylene terephthalate (PET), and paper.

Practical applications are illustrated across healthcare, environmental monitoring, and industrial automation domains. Examples include wearable printed thermistor patches for continuous body temperature monitoring in medical contexts, printed humidity sensors used for environmental climate control and air-quality monitoring, and flexible distributed sensor networks integrated within industrial IoT systems.

The integration of artificial intelligence (AI) and machine learning (ML) techniques within the sensor data workflow is also examined, demonstrating how intelligent algorithms enhance sensor calibration, compensate for drift, and enable predictive analytics. Such AI-driven data processing significantly improves the quality of sensor-derived insights, including enhanced accuracy in environmental forecasting and the early detection of anomalies.

Finally, practical workflows and design considerations critical to developing robust printed sensors are presented. The comprehensive process, from material selection and ink formulation through device patterning, printing, and curing optimization, to encapsulation, calibration, and reliability testing, is detailed. These guidelines aim to support researchers and engineers in developing flexible printed temperature and humidity sensors that meet the stringent performance and stability requirements of real-world applications.

2. Principles and types of printed temperature and humidity sensors

Understanding the physical principles behind temperature and humidity sensing is fundamental to the development of advanced sensor technologies that form the backbone of modern environmental monitoring, industrial process control, and emerging IoT applications. The measurement of these parameters relies on carefully engineered materials that exhibit predictable and reproducible responses to thermal and hygroscopic stimuli, converting these physical changes into quantifiable electrical signals. This section examines the fundamental physical principles governing temperature and humidity sensing mechanisms, providing a comprehensive analysis of different sensor types, their underlying physics, material science considerations, and the unique advantages offered by printed electronics implementation.

3. Temperature sensors

Temperature sensing relies on materials whose physical properties exhibit well-defined, predictable relationships with temperature. The central challenge is to convert thermal energy into a measurable electrical signal while maintaining accuracy, stability, and linearity across the operating range. **Figure 1** provides schematic examples of common temperature-sensor architectures. The following sections examine resistive temperature detectors (RTDs), thermistors, thermocouples, and semiconductor, thermoelectric, and optical sensors, outlining their operating principles, strengths, and application limits.

Resistive temperature detectors represent the most accurate and stable class of temperature sensors for industrial applications, utilizing the fundamental relationship between electrical resistance and temperature in pure metals [11]. The physical principle underlying RTD operation is based on the temperature dependence of electrical resistivity in metallic conductors, where increased thermal energy enhances phonon scattering of charge carriers, resulting in increased

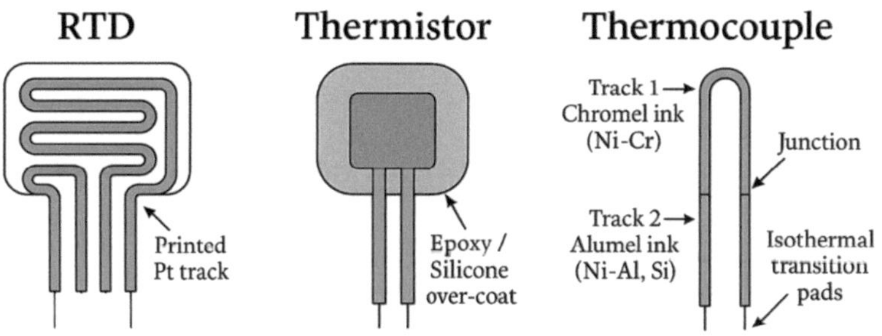

Figure 1.
Schematic comparison of three printed temperature-sensor architectures. The left diagram shows a printed platinum track representing a resistance temperature detector (RTD). The center diagram shows a thermistor composed of a rectangular sensing element with an epoxy or silicone overcoat and two electrical contacts. The right diagram shows a printed thermocouple with two conductive tracks: Chromel ink (Ni–Cr) and Alumel ink (Ni–Al, Si), joining at a junction and connected to isothermal transition pads.

electrical resistance [12]. This phenomenon follows a predictable, nearly linear relationship described by the Callendar-Van Dusen equation for platinum RTDs, which accounts for both linear and higher-order temperature coefficients [13].

The construction of RTD sensors involves two primary approaches: wire-wound and thin-film configurations. Wire-wound RTDs utilize fine platinum wire, typically 0.1–0.5mm diameter, wound around a ceramic or glass former, providing excellent accuracy and stability but with slower response times due to increased thermal mass [14].

The integration of RTDs into printed electronics platforms has been demonstrated through various approaches, including screen-printed platinum pastes and inkjet-printed platinum nanoparticle inks [15]. Printed RTDs maintain the fundamental sensing mechanism while offering advantages such as flexible substrates, large-area coverage, and reduced manufacturing costs. However, challenges include achieving the purity and stability of platinum films comparable to bulk RTDs and managing the effects of substrate thermal properties on sensor performance [16].

On the other hand, thermocouples operate on the Seebeck effect, which generates an electromotive force (EMF) at the junction of two dissimilar metals when subjected to a temperature gradient. The physical mechanism involves differences in Fermi energy levels and charge carrier densities between materials, creating a potential difference proportional to the temperature difference between the measurement and reference junctions. The Seebeck coefficient quantifies the voltage generated per unit temperature difference, with common types including K-type (41 $\mu V\ {}^{\circ}C^{-1}$), J-type (52 $\mu V\ {}^{\circ}C^{-1}$), and T-type (43 $\mu V\ {}^{\circ}C^{-1}$), each optimized for specific temperature ranges and applications [17].

Thermocouples present challenges, including low output voltages requiring precise amplification, nonlinear response characteristics, and the need for cold junction compensation. Material selection involves balancing Seebeck coefficients, temperature stability, and oxidation resistance, with noble metal types offering superior high-temperature performance at higher costs compared to base metal alternatives [18].

In printed electronics applications, thermocouples are fabricated using screen printing or inkjet printing with conductive inks containing different metals or alloys. The critical challenge involves achieving precise composition control and junction quality necessary for stable thermoelectric properties. Recent developments have demonstrated printed thermocouples using silver [19] and carbon-based inks [20, 21], though Seebeck coefficients remain lower than those of conventional metal combinations, requiring optimization of ink formulations and printing parameters to enhance performance.

Regarding thermistors, these devices follow an exponential resistance–temperature relationship, and negative temperature coefficient variants, composed of manganese, nickel, and cobalt oxides, provide high sensitivity at the expense of nonlinearity, often requiring Steinhart–Hart or similar correction models [22]. In a related but distinct category, the same underlying principles have been adapted to printed electronics, where semiconductor inks based on nanoparticles, organic compounds, carbon nanotubes, or graphene are used to create flexible films with tunable temperature coefficients [23–25]. In such systems, achieving long-term stability and accuracy remains a central challenge, as it depends on carefully controlling the relationship between ink formulation, processing conditions, microstructure, and electrical percolation.

Beyond conventional sensing mechanisms, specialized temperature sensors utilize advanced physical principles, including enhanced thermoelectric effects and optical phenomena. Thermoelectric sensors, based on the Seebeck effect, can be enhanced through nanostructuring and material engineering to achieve higher sensitivity and faster response times. Recent developments in flexible thermoelectric materials, including organic conductors and semiconductor nanowires, have enabled printed thermoelectric temperature sensors with improved performance characteristics.

Optical temperature sensors utilize temperature-dependent optical properties such as fluorescence lifetime, absorption spectra, or refractive index changes [26, 27]. These sensors offer electromagnetic interference immunity, high sensitivity, and remote sensing capabilities [28, 29]. However, their printed electronics implementation remains challenging due to the requirements for precise optical components and integrated light sources.

Piezoelectric temperature sensors represent another specialized approach, exploiting the temperature dependence of piezoelectric coefficients in materials such as quartz, lithium niobate, and gallium orthophosphate [30, 31]. These sensors prove particularly valuable for high-temperature applications where conventional sensors fail, with some piezoelectric materials remaining stable above 1000 °C [32]. However, integration into printed electronics platforms remains limited due to the specific crystal structures and specialized processing requirements of piezoelectric materials, making them less suitable for solution-based printing techniques compared to other sensor technologies [33, 34].

4. Humidity sensors

Humidity sensing relies on the interaction between water-vapor molecules and sensing materials, which induces measurable changes in properties such as electrical conductivity, dielectric constant, or optical response. The central challenge is to ensure a selective response to water vapor while maintaining stability against interfering species and environmental fluctuations [1]. A representative capacitive configuration featuring interdigitated electrodes located below a porous sensing

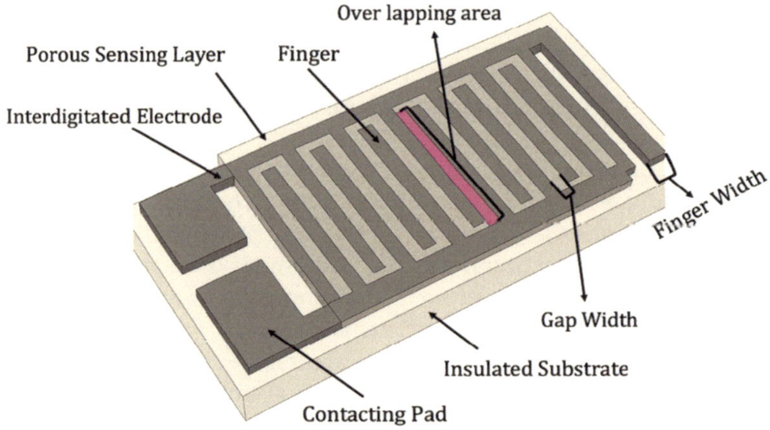

Figure 2.
Schematic 3D representation of a capacitive humidity sensor showing an insulated substrate with interdigitated electrodes, a porous sensing layer covering the active area, and contacting pads for electrical connection. The geometry includes alternating finger and gap widths defining the sensing region.

layer on an insulating substrate is shown in **Figure 2**, illustrating the geometry typically used to transduce humidity-induced dielectric changes into an electrical signal.

One category comprises capacitive humidity sensors, whose operation relies on the change in the dielectric constant of a hygroscopic medium when exposed to water vapor. Water molecules are adsorbed within the sensing film placed between or across electrodes, so the effective permittivity rises because liquid water has a dielectric constant of about 80, whereas typical polymer dielectrics fall in the range of about 2 to 15, resulting in a measurable increase in capacitance. These devices are commonly implemented as parallel-plate stacks or as interdigitated electrodes beneath a porous sensing layer on an insulating substrate, which promotes rapid vapor access and short diffusion paths [35].

The transduction follows $C = \varepsilon_0 \varepsilon_r A/d$, where ε_0 is the vacuum permittivity, ε_r is the effective relative permittivity of the hygroscopic layer, A is the effective electrode overlap area, and d is the dielectric thickness or electrode gap. As moisture penetrates the film, ε_r increases according to effective-medium or mixing descriptions that depend on the water volume fraction and its spatial distribution within the polymer matrix, as well as the porosity and morphology imposed by the chosen materials and processing conditions [36].

Performance characteristics include excellent linearity across full relative humidity ranges (0–100% RH) [37], good long-term stability [38, 39], and relatively fast response times between 10 and 30 s [40, 41]. However, limitations include temperature dependence, sensitivity to electromagnetic interference, and potential contamination from volatile organic compounds.

Resistive humidity sensors detect moisture through electrical conductivity or impedance changes in hygroscopic materials. The fundamental mechanism involves water vapor adsorption onto surfaces or into bulk sensing materials, enhancing ionic conductivity through water molecule dissociation and hydrated ion formation. This occurs through two primary mechanisms: ionic conduction via adsorbed water layers and electronic conduction changes in semiconductor materials [42].

In ionic conduction sensors, water molecules adsorb onto ceramic or polymer surfaces containing mobile ions. Adsorbed water dissociates according to $H_2O \rightleftharpoons H^+ + OH^-$, creating charge carriers that facilitate ionic transport [36]. The conductivity increases correlate with water activity and ionic species concentration in the sensing material. Common materials include ceramic oxides (Al_2O_3, TiO_2, ZnO) [43, 44], salt-polymer composites [45], and polyelectrolyte systems [46].

Beyond electrical transduction, some humidity sensors exploit changes in optical properties, such as absorption, fluorescence, or refractive index, which are induced by water vapor interactions with specific materials [3]. These optical approaches include sensors that monitor shifts in absorption or emission spectra, detect color changes in chromogenic materials or photonic crystals, or utilize the swelling of hygroscopic films to alter light transmission. Recent advances in solution-processable materials, including perovskites, are opening new possibilities for fully printed optical humidity sensors [47–49].

In addition to optical methods, thermal conductivity humidity sensors determine absolute humidity by comparing the thermal conductivities of dry and humid air [50]. This approach relies on the distinct thermal properties of water vapor and typically employs matched thermal elements arranged in bridge configurations [51]. While printed heater elements and temperature sensors could simplify implementation, achieving reliable thermal management and mechanical stability remains complex in printed formats [15].

A further strategy is provided by gravimetric sensors, which measure humidity by tracking mass changes as water vapor adsorbs onto sensitive materials [52]. Resonant structures, such as quartz crystal microbalances, detect frequency shifts caused by mass loading [53, 54]. Although this technique offers exceptional sensitivity [55], its adoption in printed electronics is limited by the demands of precise frequency measurement and mechanical robustness. Nonetheless, printed resonators and flexible substrates are enabling promising proof-of-concept devices [56].

5. Background and state of the art

Printed flexible sensors for temperature and humidity have advanced significantly over the past decade, transitioning from laboratory prototypes to practical real-world applications. Unlike traditional rigid sensors based on silicon that require complex microfabrication, printed sensors are manufactured through additive techniques such as screen printing, inkjet, gravure, and aerosol jet printing (AJP). These methods deposit functional inks, including conductive, dielectric, or sensing materials, onto flexible substrates in precise patterns, enabling low-cost mass production with minimal material waste. The capacity to print on large-area, lightweight plastics, paper, or textiles facilitates the creation of inexpensive, disposable sensors that conform easily to diverse surfaces. A comparative summary of recent developments in printed flexible humidity sensors is presented in **Table 1**, while **Table 2** provides an overview of recent advances in printed flexible temperature sensors.

Over the last decade, printed humidity and temperature sensing has progressed from thick-film feasibility on rigid substrates to application-oriented devices on flexible and even biodegradable carriers. Screen printing first established reliable

Printing technique	Sensor type	Substrate	Ink/Functional material	Post-treatment	Printing resolution*
Screen printing [57]	Resistive RH sensing	96.6% alumina (Al_2O_3)	1) Au resinates for IDE; 2) water-soluble phthalocyanine active layer; 3) polystyrene protective film	Gold firing 850°C/15 min/oven; organic layers dried <100°C	40–50 µm line/space, Au film ≈ 0.34 µm
Screen printing [58]	Flexible impedance-type RH sensor for smart garments	Polycotton fabric	Nafion® 117 proton-conducting film (<1 µm)	Ag cure 130 °C/10 min; Nafion dry 60°C/1h + 168 h RT; interface UV.	IDE fingers 1 mm; gap 1–3 mm
Roll-to-roll gravure [59]	Humidity sensing for smart packaging	PET (75 µm)	Conducting pol ymer ink (PH-001, Paru Co.)	150 °C/5 s/ heat curing	Film thickness ≈ 400 nm
Ink-jet printing [60]	Ultra-sensitive capacitive RH sensor	PET film	Monolayer graphene oxide or few-layer black-phos-phorus flakes	Ag sinter 120°C/30 min; GO/BP dried 80°C	Ink-jet line ≈ 100 µm; IDE pitch ≈ 500 µm
Screen printing [41]	Flexible capacitive RH sensor	PET (Melinex ST504)	Conductive ITO microparticle ink (plates); Al_2O_3 dielectric ink (3 layers)	110°C/10 min each layer	230-mesh screen; 1 mm finger width (geometric)
Ink-jet printing [61]	Low-RH FET humidity sensor	Si/SiO_2 FET die	Pt-doped In_2O_3 NP ink	300°C/2 h/ air bake	Drop-on-demand; spot pitch ≈ 30 µm
Screen & flexographic print [35]	Capacitive RH sensor	PET film	Ag NP ink (bot-tom & grid top); cellulose-acetate -butyrate (CAB) sensing dielectric	Ag cure 60°C/10 min; CAB dried 90°C	Grid: 200 µm line & opening; 55 µm AJP lines demonstrated
Aerosol jet printing [83]	Flexible resis-tive RH sensor; respiration/IoT	PET film	Triton-X100-exfoliated gra-phene ink + gelatin binder (0–1 mg mL^{-1})	Stage 100°C during print; film incubation 40°C/1 h	IDE gap 100 µm; sensing stripe 1 mm × 0.5 mm (film 400 nm)
Aerosol spray [62]	High-sensitivity resistive RH sensor	PET with sputtered Au/Cr IDEs	Sonication-exfo-liated MoS_2 nanoflake ink (water:IPA 7:3)	Substrate held at 80°C during spray; no further cure	IDE photolitho-graphic; film area 50 mm × 4 mm
Inkjet printing [63]	Capacitive RH sensor for breath monitoring/ wearable mask	Glossy ink-jet photo paper (180 gsm)	Commercial graphene ink (water-based)	Air-dry; no thermal cure reported	Finger width 1 mm; finger spacing 1 mm; total area 16 × 25 mm^2
Screen printing [64, 65]	Flexible resis-tive RH sensor	Glossy photo paper (150 gsm)	Graphene-carbon conduc-tive ink; porous PDMS overcoat	60°C/15 min dry	IDE gap 750 µm; line 750 µm; layer 20–25 µm

Printing technique	Sensor type	Substrate	Ink/Functional material	Post-treatment	Printing resolution*
Screen printing [65]	Fast resistive RH sensor	PET film	Hexagonal-WO_3 nanowires; Ag IDE	60°C/5 min (ink dry)	IDE finger/space 50 μm
Aerosol jet printing [84]	Resistive sensors for IoT applications; gas sensing (N_2, H_2S, NH_3)	PET film	Graphene inks with dispersants: gelatin (GGe), triton X-100 (GTr), tween-20 (GTw)	40°C/overnight/incubation; 70°C stage during printing	300 μm nozzle; ≈ 10 μm printing capability; 150 μm gap spacing
Aerosol jet printing [85]	Resistive temperature detectors (RTDs)	Polyimide (Kapton) film (125 μm thickness)	Silver nanoparticle ink (JA-A426, 50 wt% AgNPs, aqueous)	260°C/1 h/ambient atmosphere (thermal sintering)	Line width ≈ 70 μm; trace thickness 0.15–1.5 μm (1–6 layers)
Pencil drawing [67]	Flexible resistive RH sensor; human breath monitoring	Conventional A4 printing paper (cellulose fibers)	2B pencil graphite + ZnO nanoparticle paste in deionized water	ZnO paste applied via doctor blade; dried at room temperature	Bar device design; uniform coating via 10 pencil strokes
Electrohydrodynamic inkjet printing [66]	Capacitive RH sensor for in-space manufacturing	Polyimide (PI) film	Silver nanoparticle ink (50 wt %) + La-doped $BaTiO_3$ ink (35 wt% $BaTiO_3$, 1 wt% KOH, surfactants)	Annealed at 250°C for 10 min	100 μm glass nozzle; 50 μm finger spacing, 30 μm electrode width, 2 mm finger length
Screen printing [68]	Ultrathin (6 μm) capacitive RH sensor for leaves	Ethyl-cellulose film	Carbon microparticle ink; EtC encapsulation	Cure 100°C/30 min; EtC water-swelled transfer	IDE finger ≈ 500 μm; total area 1 cm^2
Screen printing [69]	Wearable sweat-rate/resistive RH element	Polyimide (Kapton)	Conductive carbon-black ink; Ag tracks	Ambient dry 30 min; optional 100°C/10 min	230-mesh screen; resistor 8.75 mm × 0.5–2.5 mm; thickness 5–9 μm
Screen printing [70]	Flexible resistive RH sensor	Polyimide (Kapton)	3C-SiC nanoparticle ink (40 wt% in SOL725); Ag paste IDE	Ag cure 300°C/60 min; SiC cure 300°C/180 min	325-mesh; IDE line 0.5 mm/gap 0.5 mm; SiC disk Ø 7.5 mm (23 μm thick)
Aerosol jet printing [86]	Multifunctional humidity and temperature flexible sensor	Chitosan (CS) flexible film	$Ti_3C_2T_x$ MXene + sodium alginate (SA) polyelectrolyte ink (0.5 wt% MXene)	None (room temperature printing); 10 printed layers	Line width ≈ 55 μm (350 μm nozzle); printing speed 2 mm s^{-1}

Table 1.
Comparative summary of printed humidity sensors reported in the literature.

Printing technique	Sensor type	Ink/Functional material	Substrate	Post-treatment	Printing resolution
Screen printing [71]	PTC/NTC Thermistor – flexible probe	Polymeric PTC, NTC oxide Sn–Sb + Au 30 nm	PEN 125 μm	130°C/30 min + O_2 plasma + CYTOP 100°C/20 min	PTC ≈ 8 μm; NTC ≈ 10 μm
Inkjet printing [72]	Thermistor for body temp	Ag complex (TEC-IJ-010)	Kapton HN	130°C/10 min + 150°C/30 min	Line 300 μm/ pitch 60 μm/ thick 0.37 μm
Inkjet + screen printing [73]	Skin-conformable temp sensor	Graphene/PEDOT: PSS (1–5 wt% each) + stretchable Ag	Polyurethane bandage	130°C/10 min + optional Novec 1,700 coating	20 μm drop spacing, 3 layers, 1270 dpi
Aerosol jet printing [74]	Mini NiO thermistor, 30–250°C	NiO NP ink (0.5 g total)	Al_2O_3	Ag 180°C/0.5 h + NiO 200°C/1 h	≈0.1 mm sensor
Direct writing [75]	Thick-film Pt RTD, 25–500° C	Pt paste ESL 5545 (50 wt%)	96% alumina	Pt 1200°C/15 min + Ag 650°C/15 min (air)	Lines 145 μm/ gap 75 μm
Air-spray coating [76]	Flexible r-GO sensor	r-GO, SWCNT, MWCNT solutions	PET 100 μm + O_2 plasma	110°C during coating + tape 500 N	Mask plate control + 10s spray
Screen printing [93]	Flexible RTD	Ni–NP thermoplastic	Kapton	135°C/5 min	Line 200 μm/ pitch 400 μm/ 17 μm thick
Inkjet printing [77]	Temp/ Humidity 20–70°C	PEDOT:PSS (1.3 wt%)	PET	60°C/10 min	150–200 μm footprint
R2R gravure [78]	NFC tag temp thermistor	Ag–NP hydrophobic + glass frit	PI 75 μm	250°C/20 min	Line 0.7 mm/ pitch 0.3 mm/ thick ≈ 0.6 μm
Inkjet printing [79]	Flexible thermistor	Ag NP (homemade vs. comm.)	PET 107 μm	120°C/10 min	500 μm line; 0.4 mm gap
Inkjet printing [80]	Wearable temp sensor	PEDOT:PSS + cross-linker/GOPS/TX-100	PEN 50 μm	140°C/30 min + CYTOP 140°C/30 min	≈1 μm sensing layer
Aerosol jet printing [81]	NiO NTC thermistor	NiO NP ink (0.5 g/10 g)	Kapton tape	UV 60% 10 min or 260°C/1 h	150 μm nozzle
Screen + Aerosol jet printing [82]	Flexible thermistor	Ag NP (electrodes) + C-black NTC ink	PEN 100 μm	Ag flash 420–480 V, 200 μs; poly 120°C/6 h	60 μm pixel; pitch 500 μm array
Inkjet printing [16]	IEC-compliant temp sensors	Ag NP ink I30EG-1	LCP + COC	200°C/24 h + FC-742 coating	1270 dpi
Aerosol jet printing [85]	Flexible RTD	Ag NP 50 wt%	Kapton	260°C/1 h; PI passivation 5 μm	≈70 μm line; 0.15–1.5 μm thick

Printing technique	Sensor type	Ink/Functional material	Substrate	Post-treatment	Printing resolution
Screen printing [15]	NTC SiC	β-SiC NP 30 wt% in PI	Kapton	200°C/60 min	150–200 μm
Inkjet printing [94]	Capacitive, 100–300°C	Ag NP; hybrid SiOx-polymer	PI 50 μm	150°C/30 min	Drop-spacing 20 μm
Aerosol jet printing [95]	Capacitive, 22–80°C	PVAC 5 mg mL^{-1} + PCL (2 mg mL^{-1})	Glass	PCL UV crosslink + 120°C	150 μm nozzle
Inkjet printing [90]	Graphene thermistor	Nanocomponent graphene/ethyl cell.	PDMS	Xenon flash 3kV (≤14 pulses)	100 μm drop; film 0.53 μm thick
Screen printing [87]	Thermocouple	p–$Bi_{0.5}Sb_{1.5}Te_3$/ n–$Bi_2Te_{2.7}Se_{0.3}$	PI	Flash cure ms	leg 0.7 mm
Screen printing [88]	Thick-film W–Re thermo-couple 1600°C	W, WRe26 powder 1 μm	Al_2O_3	Sinter Ar 1500°C/ 30 min + anneal 300°C/6 h	≈30 μm thick film
Screen printing [91]	NTC Thermistor, 30–170°C	β-SiC NP (β-SiC)	Kapton	200°C/60 min	300 μm line
Screen printing [89]	Thermocouple	ITO + In_2O_3 + Ag shield paste	Al_2O_3	300°C activation + Ag 150°C/2 h	film linewidth 4 mm
Inkjet printing [92]	Thermocouple	In_2O_3 + ITO NP + frit	Al_2O_3	500°C/1 h + 1250°C/3 h	width 0.7 mm; pitch ≈ 10 μm/ n ≈ 5 μm

Table 2.
Comparative summary of printed temperature sensors in the literature.

baselines on ceramics and textiles before higher-resolution nozzle methods such as inkjet, AJP, and electrohydrodynamic (EHD) inkjet broadened the design space and lowered thermal budgets. On alumina, Soukup et al. [57] screen-printed gold-resinate interdigitated electrodes (IDEs; 40–50 μm line/space; Au ≈ 0.34 μm) and overlaid a sub-micrometer, water-soluble phthalocyanine film. A short gold firing (850 °C for 15 min) followed by sub-100 °C drying yielded reproducible exponential impedance–RH behavior across 20–90% RH, indicating that fine-feature multilayer thick-film routes can deliver stable metrology on non-flexible dielectrics. On textiles, Kutzner et al. [58] used a UV-cured polyurethane planarization layer on poly-cotton, screen-printed Ag IDEs, and micro-dispensed Nafion® 117 (< 1 μm). The ≈ 125 μm stack remained bend-tolerant while exhibiting 60–140 s adsorption times and ≈ 15% hysteresis over 30–90% RH, quantifying the trade-off between mechanical compliance on porous substrates and reversible sorption needed for precision.

Kang et al. [59] demonstrated the web-speed manufacturability of a humidity-activated sensor signage tag using fully roll-to-roll gravure on 75 μm PET. A conducting polymer ink formed the RH-active film with ≈ 400 nm thickness, consolidated by a 150 °C/5 s heat cure. The film shows rapid water uptake and a resistance that decreases with

RH; above ≈ 70% RH, the device operates as a threshold switch that drives real-time on/off display action. Sensitivity is tunable via active layer thickness, emphasizing that, in high-throughput R2R implementations for packaging, the humidity response is chiefly engineered through the composition and thickness of the printed polymer film. He et al. [60] printed Ag IDEs (≈ 100 μm lines, 500 μm pitch) on PET and added monolayer graphene oxide or few-layer black phosphorus, dried at 80 °C. The devices covered 11–97% RH with large low-frequency capacitance changes of ≈ 4.45×10^4 for graphene oxide and ≈ 5.08×10^3 for black phosphorus at 10 Hz, and response times of 2.7–4.7 s, linking surface chemistry and nanometric thickness to high gain and fast transduction. In parallel, fully printed capacitors showed how screen printing can still deliver broad-range readouts: two-plate ITO–Al_2O_3 stacks on PET cured at 110 °C/10 min per layer achieved 5–95% RH operation, 0.85–7.76 pF $\%RH^{-1}$ sensitivity, 21 s response/4.8 s recovery, and 26–35% hysteresis, underscoring how porosity-driven sensitivity can be offset by desorption-limited reversibility [41].

From 2020 onward, additive manufacturing broadened both material choice and device physics in printed humidity sensing. Inkjet patterning of platinum-doped In_2O_3 nanoparticle channels on Si/SiO_2 yielded FET-type devices that detected low relative humidity in the 3.3–18% RH range after baking at approximately 300 °C, where the response arose from work-function modulation in sub-micrometer films [61]. In parallel, screen and flexographic printing enabled silver and cellulose acetate butyrate parallel-plate capacitors with 18–25 μm dielectrics dried at 60–90 °C. These sensors operated from 3 to 100% RH, delivered 1.0–1.5 pF % RH^{-1} sensitivity, showed response and recovery times of 31 s and 8 s, and limited hysteresis near 2.9%, confirming that hydrophilic dielectrics can suppress hysteresis without sacrificing dynamic range [35]. Complementing these routes, aerosol-sprayed MoS_2 deposited on Au/Cr interdigitated electrodes at 80 °C formed porous films about 10 layers thick, produced conductance changes on the order of 10^6, reached 8 s and 22 s response and recovery, exhibited ≈ 6.5% hysteresis, and remained stable for 35 days [62].

More recent work emphasized low thermal budgets and biodegradable carriers. One-step printing of graphene interdigitated electrodes on glossy photo paper, air-dried without thermal curing, enabled mask-embedded breath monitors with response times below 5 s and relative standard deviation at or below 2.2% [63]. Building on this, screen-printed graphene–carbon electrodes on photo paper capped with porous PDMS balanced low hysteresis around 2.9% and bendability down to a 20 mm radius while sustaining 31 s and 8 s response and recovery across 25–91% RH [64].

At the fast-response limit, screen-printed hexagonal WO_3 nanowires on PET with 50 μm finger and space, dried at 60 °C for 5 min, achieved 1.5 s response and 15 s recovery and reached sensitivities up to approximately 2.9×10^3 kΩ $\%RH^{-1}$, a performance attributed to porous networks that expose active planes [65]. Electrohydrodynamic inkjet co-printing of silver at about 50 wt% together with lanthanum-doped barium titanate on polyimide, using 100 μm nozzles with 50 μm spacing and annealing at 250 °C for 10 min, produced capacitive sensors suitable for in-space manufacturing and respiratory monitoring, with 0.12 nF $\%RH^{-1}$ sensitivity, approximately 9 s response, about 2 s recovery, and hysteresis below 5% [66]. Finally, resource-lean approaches such as pencil-drawn graphite combined with doctor-bladed ZnO on A4 paper still delivered up to roughly 717% sensitivity at 97% RH, with ≈ 4.31 s response and ≈ 10 s recovery after ambient drying [67].

By 2025, integration, biocompatibility, and reliability were central. An ultrathin conformal sensor under 6 μm, made by screen-printing carbon interdigitated electrodes on ethyl cellulose and transferring the film to living foliage, operated from 1 to 85% RH. Above roughly 55% RH, it delivered 50–1000 pF %RH^{-1} sensitivity with a baseline near 16 nF, and it maintained multi-day stability with high signal-to-noise, showing that conformal bio-interfaces can be metrologically robust [68]. Water-based, screen-printed carbon-black resistors on Kapton at 5–9 μm thickness endured 500 bends at a 1 cm radius and still produced a usable humidity response, with resistance increasing by about 58% between 35 and 85% RH [69]. Wadhwa and colleagues printed silver interdigitated electrodes followed by a 3C-SiC nanoparticle layer at 40 wt% in SOL725 on 125 μm polyimide, using 0.5 mm lines and gaps and a 7.5 mm SiC disk at 23 μm thickness; curing at 300 °C for 60 min for silver and 180 min for SiC yielded a respiration sensor for KN95 masks operating from 10 to 90% RH with $\Delta R/R_0$ near 45.2%, sensitivity around 5.34 Ω %RH^{-1}, adsorption and desorption times of 18 s and 46 s, and hysteresis close to 6.5% at 60% RH, stable for approximately 3.5 h [70]. Key studies and fabrication details for printed humidity sensors are summarized in **Table 1.**

Printed temperature sensing follows a parallel trajectory shaped by accuracy and calibration requirements. Screen-printed polymeric PTC and NTC layers, about 8–10 μm thick, on polyethylene naphthalate, encapsulated with O_2 plasma and CYTOP, produced flexible 15 × 15 cm probes that operated from 20 to 80 °C with a sensitivity near 5 mV $°C^{-1}$ at 4.8 V and sub-millisecond dynamics, indicating that thick-film routes can meet basic thermometry on polymer foils [71]. Flexibility and on-skin operation were emphasized in two studies. Inkjet-printed silver on Kapton used roughly 300 μm traces and 130–150 °C anneals, achieving a TCR of about 2.2×10^{-3} $°C^{-1}$ over 20–60 °C with hysteresis below 5% [72]. A second design combined inkjet-printed graphene and PEDOT:PSS with screen-printed stretchable silver on polyurethane bandages at 1270 dpi and 20 μm drop spacing, reaching TCR values between 0.047 and 0.064% $°C^{-1}$ and responses of about 20 s for heating and 18 s for cooling [73].

Between 2017 and 2019, the materials palette and operating envelope widened. Aerosol-jet-printed NiO NTCs on alumina with silver lines near 80 μm, followed by 180–200 °C post-treatments, delivered response times under 1 s up to 250 °C [74]. Extrusion-printed thick-film platinum RTDs on 96% alumina, fired at 1200 °C for 15 min, exhibited a TCR near 3800 ppm K^{-1} and ± 0.2 °C accuracy to roughly 350 °C [75]. Air-sprayed reduced graphene oxide on PET provided a TCR around 0.6345% $°C^{-1}$ with a response near 1.2 s [76]. Consumer-grade inkjet printing of PEDOT:PSS on PET with silver interdigitated electrodes achieved a TCR near − 0.8% $°C^{-1}$ and linearity with R^2 greater than 0.99 over 20–70 °C [77].

By 2020, the focus shifted to throughput and integration. Roll-to-roll gravure on 75 μm polyimide incorporated thermistors into NFC tags after a 250 °C, 20 min cure, operating from − 10 to 30 °C with about 1 °C resolution and response times near 5–6 min [78]. Comparative inkjet studies of silver showed linear behavior from 30 to 100 °C with R^2 greater than 0.99 and a sensitivity around 0.108 Ω $°C^{-1}$ after 120 °C sintering [79]. Co-printed PEDOT:PSS with GOPS and Triton X-100, paired with silver electrodes on PEN and sealed with CYTOP, yielded wearable sensors stable between 30 and 80% RH, with responses of roughly 1.5 s for heating and 6 s for cooling across 25–50 °C, illustrating how encapsulation and crosslinking limit humidity-induced drift [80].

In 2022–2023, arrays and post-treatments pushed performance: AJP-printed five-layer NiO NTCs on Kapton benefited from UV curing that reduced process time by ≈ 90% while sustaining 100–180 °C operation [81]. A 25 × 25 pixel NTC matrix on PEN, with pixels around 60 μm and 500 μm pitch, and with flash-cured silver at

420–480 V for 200 μs plus 120 °C polymer bakes, delivered sub-second pixel responses and an average error below 0.5 °C across 5–90 °C [82]. Inkjet sensors using in-plane IEC-oriented silver on liquid-crystal polymer and cyclic olefin copolymer reported TCR values of about 1.23–1.31 × 10^{-3} K^{-1} with hysteresis under 1.4% from 10 to 85 °C [16].

From 2023 to 2025, three thrusts stood out. High-sensitivity metals were demonstrated by aerosol-jet-printed silver RTDs on Kapton with a TCR of about 3.7 × 10^{-3} °C^{-1}, a time constant near 0.83 s, and a linear response from – 20 to 150 °C [85]. Ceramic–polymer composites, such as β-SiC in polyimide, reached a TCR of about – 0.556% °C^{-1} and ± 1.35 °C error over – 25 to 170 °C with acceptable bending drift [15]. Ultrafast and thermoelectric routes advanced through screen-printed p–$Bi_{0.5}Sb_{1.5}Te_3$ and n–$Bi_2Te_{2.7}Se_{0.3}$ legs on polyimide with photonic flash curing, yielding sensitivities of about 124–319 μV °C^{-1}, millisecond responses, and power factors above 10 μW cm^{-1} K^{-2} [87]. W–Re thick-film thermocouples sintered at 1,500 °C operated from roughly 100 to 1,600 °C with ±1.5% accuracy and a response time near 1.74 s [88]. ITO–In_2O_3 thermocouples with silver shielding added about 43.68 dB of electromagnetic protection and achieved a Seebeck coefficient near 20.87 μV °C^{-1} [89]. Inkjet graphene with xenon-flash photonic sintering achieved a TCR of around 1.19 × 10^{-3} °C^{-1}, transients near 53 μs, and bending endurance at about a 20 mm radius for 500 cycles [90].

In parallel, Wadhwa and colleagues screen-printed β-SiC nanoparticles dispersed in polyimide at 30–40 wt% onto Kapton using silver interdigitated electrodes about 300 μm wide, cured at 200 °C for 60 min, and obtained an NTC thermistor operating from – 30 to 170 °C. At 30 wt% loading, the composite showed B near 5.02 × 10^2 K and sensitivity around – 0.56% °C^{-1} over 25–170 °C, with humidity-induced drift close to 6.5% [91]. At the high-temperature limit, Miao and co-workers inkjet-deposited In_2O_3/ITO thermocouples on alumina and used two-stage consolidation at 500 °C for 1 h, followed by 1,250 °C for 3 h, yielding dense films stable to about 1,600 °C. The devices reported a Seebeck coefficient near 58 μV °C^{-1} over – 20 to 1,600 °C, response around 53 μs, and drift near 6.6 °C h^{-1} under the reported conditions [92].

Taken together, screen and gravure remain well-suited for area and layer count but often require higher-temperature consolidation; inkjet, AJP, and EHD provide fine features, short thermal transients, and compatibility with polymers at modest cures. Encapsulation and crosslinking explicitly address humidity cross-sensitivity in thermal devices (e.g., CYTOP-sealed PEDOT:PSS), while ceramic–polymer systems such as β-SiC composites illustrate the remaining drift that must be managed for long-term accuracy. Arrays now commonly reach <0.5 °C error with <1 s per-pixel dynamics, and EMI-shielded thermocouples show that printed stacks can be engineered for electrically noisy environments.

Beyond device-level performance, integration and system aspects increasingly shape design choices. Conformal interfaces, such as ultrathin ethyl cellulose pathers on biological surfaces, indicate that host–sensor mechanics and surface chemistry can be addressed without sacrificing metrology, provided the polymer carrier, electrode geometry, and film thickness are co-designed. Power budgets remain underreported across the literature; while crosslinked and encapsulated PEDOT:PSS thermistors have demonstrated wearable monitoring with ≈ 1.5 s/ ≈ 6 s dynamics, comprehensive energy analyses, including duty cycling, sampling strategies, and rectification/conversion losses, would enable more realistic assessments for IoT and battery-free nodes.

Sustainability and scalability considerations span the spectrum from ambient-processed, biodegradable routes to energy-intensive, high-temperature stacks. The selection of materials and processes, therefore, aligns performance targets to energy and materials footprints; a useful way forward is to report not only sensing metrics but also process energy, solvent usage, and recyclability when biodegradable carriers or water-based inks are employed.

Industrial readiness further depends on manufacturing routes and interface control. R2R demonstrations substantiate throughput, yet the most advanced material systems often rely on tightly controlled atmospheres, delicate interfaces, or multi-step sequences that can be difficult to translate into continuous lines. Simplifying post-processing, standardizing surface treatments (planarization layers on textiles, adhesion promoters on polymers), and designing inks for wider process windows can reduce sensitivity to equipment and environment.

Standardization and benchmarking would also improve comparability. Some works already align with IEC-style reporting for temperature sensors on polymer dielectrics, but many studies still use bespoke metrics or omit key test conditions. For capacitive humidity sensors, stating frequency and drive amplitude; for resistive devices, biasing and airflow/hydration conditions; and for thermocouples/RTDs, calibration windows, linearity (R^2), uncertainty, and drift over time would enable clearer cross-study evaluation and facilitate technology transfer.

Reliability over realistic lifetimes remains a central need. Published bend tests, for example, ≈ 500 cycles at 1 cm radius for carbon black films on PI and ≈ 500 cycles at r ≈ 20 mm for photonic-sintered graphene, are promising but still fall short of the millions of cycles encountered in continuous-use wearables. Long-horizon studies combining humidity/temperature cycling, mechanical strain, and contamination would better characterize calibration drift, hysteresis evolution, and end-of-life behavior. Packaging and encapsulation are as decisive as ink chemistry here: porous film control to modulate hysteresis, barrier layers to limit water uptake, and breathable encapsulants that preserve response time without introducing new cross-sensitivities are likely to have an impact comparable to changing the active material itself.

Looking forward, priorities emerge across materials, processes, and systems. On the materials side, compositions that sustain sensitivity and stability under mild processing (≤ 150–250 °C) while tolerating humidity and temperature cycling are attractive; layered and nanostructured materials show potential but would benefit from robust surface chemistry control and standardized passivation. On the process side, integrating multiple printing methods while reducing sequence complexity through lower temperature densification, in-line curing, and inks tolerant to ambient variation can bridge laboratory and R2R lines. At the system level, co-designing sensors with antennas, rectifiers, and power management, and reporting explicit energy budgets, will align developments with IoT use cases. Finally, community-adopted protocols for calibration, environmental testing, and reliability benchmarking will support the transition from isolated demonstrations to calibrated, field-deployable printed sensor systems.

6. Selection of materials

The rapidly evolving landscape of sensor manufacturing has undergone a clear shift toward additive manufacturing, particularly for temperature and humidity sensors in diverse industrial applications. This transition from traditional microfabrication to printed electronics reflects the convergence of advances in materials science, process optimization, and the growing demand for cost-effective, flexible, and application-specific sensing solutions. The inherent advantages of printed sensors, including large-area coverage, mechanical flexibility, reduced production costs, and design customization, position these technologies as enabling platforms for next-generation sensing across multiple sectors [1].

Industrial deployment of temperature and humidity sensors requires a precise mapping of sector-specific requirements that drive both material selection and manufacturing routes. As shown in **Figure 3**, the temperature-sensor network diagram highlights eight critical industrial domains, each with distinct technical challenges and operational demands. These requirements vary widely: agriculture calls for extreme environmental robustness, medical devices must meet biocompatibility standards, automotive systems demand high-temperature accuracy, and food processing environments enforce strict hygiene protocols.

Material selection for printed sensors should not begin with a catalog of inks or substrates. It should start by defining the application's critical variables: target accuracy, operating temperature and duration, response time, mechanical support,

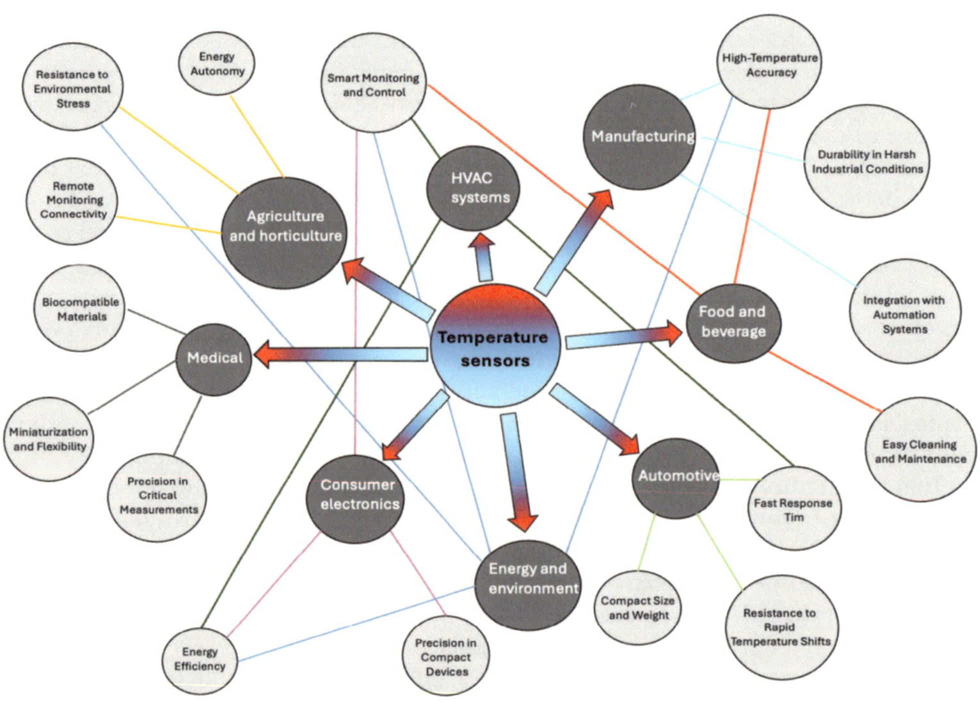

Figure 3.
Diagram showing temperature sensors at the center connected to key sectors such as automotive, consumer electronics, energy, manufacturing, HVAC, food, agriculture, and medical, with arrows linking to their specific requirements including durability, precision, and compact design.

DOI: http://dx.doi.org/10.5772/intechopen.1012577

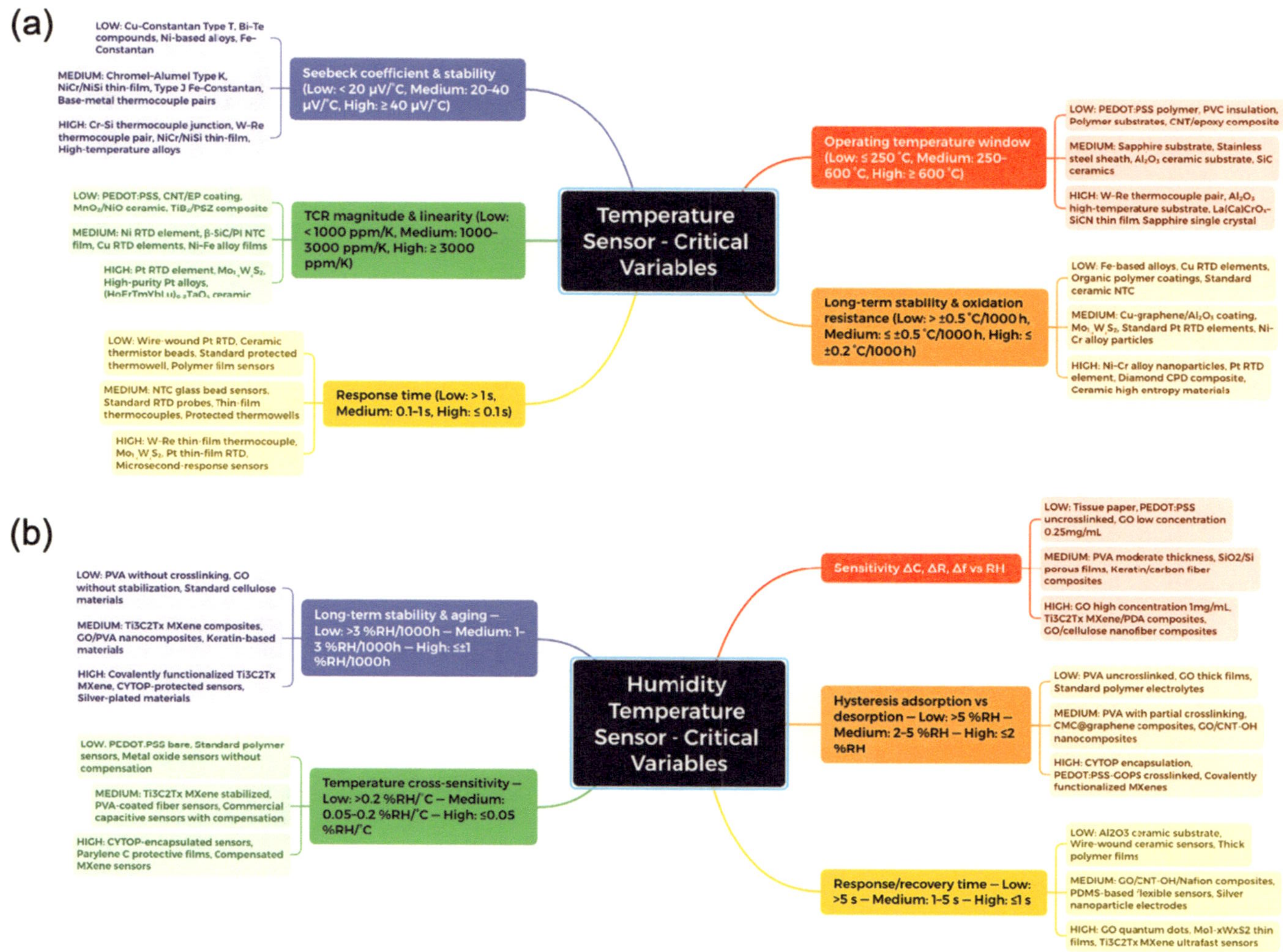

Figure 4.
Comparative diagrams summarizing critical variables for temperature and humidity sensors. Panel (a) shows temperature sensor materials classified by Seebeck coefficient, TCR magnitude, response time, operating temperature, and stability. Panel (b) shows humidity sensor materials categorized by long-term stability, sensitivity, hysteresis, temperature cross-sensitivity, and response time.

expected levels of electromagnetic interference, and any biocompatibility constraints. Once these thresholds are set for a given industry, the choices of sensing material, substrate, and protective layers should be aligned with the required performance level for each variable. The following sections translate this logic for some materials, first for temperature sensors and then for humidity sensors (**Figure 4**).

Building on the selection logic above, temperature-sensor materials are chosen based on the required TCR magnitude and linearity, the allowable thermal budget, and the target cost. When a high and linear positive TCR is the priority, noble-metal RTDs, especially high-purity platinum, are preferred and deliver ≈ 3,926 ppm K^{-1} with excellent stability [96]. For intermediate needs, sputtered nickel RTDs reach ≈ 3,000 ppm K^{-1}, while printed β-SiC/polyimide thermistors provide negative TCRs around - 5,560 ppm K^{-1} where an NTC response is acceptable [97]. In cost-sensitive or disposable designs, stabilized PEDOT:PSS films offer about - 7,700 ppm K^{-1}, trading some linearity for low processing temperature and flexibility [80].

For thermocouples, the governing metric is the Seebeck coefficient and its stability over temperature. High-temperature furnace and metallurgy environments favor tungsten–rhenium pairs with ≈ 16–28 μV K^{-1} up to 2,300 °C [98]. Chromium–silicon junctions reach at least ≈ 50 μV K^{-1} at 1,200 °C [99]. In general industrial use, type-K Chromel–Alumel and type-J iron–Constantan thermocouples provide robust performance with typical Seebeck values near 41 μV K^{-1} and 50 μV K^{-1}, respectively [100]. For field and moderate-range measurements, copper–Constantan pairs at ≈ 40–58 μV K^{-1} are practical when drift and gradient effects are controlled by calibration and junction design [101].

Architecture then follows the thermal environment. At ≥ 600 °C, W-5Re/W-26Re thermocouples remain stable up to ≈ 2,300 °C on alumina or sapphire, and La(Ca) CrO_3–SiCN films enable NTC operation from ≈ 200 to 1,100 °C with limited variation [102]. In the 250–600 °C band, metallized sapphire and stainless-steel or Inconel sheaths preserve electrical and mechanical integrity, and SiC-based thin films can support near-linear RTD behavior up to ≈ 850 °C [103, 104]. Below 250 °C, technical polymers allow fully printed devices; PEDOT:PSS is stable roughly from 30 to 80 °C, and CNT/epoxy coatings maintain a near-linear TCR from ≈ 20 to 120 °C [105].

Long-term stability and oxidation resistance depend on both materials and encapsulation. Platinum RTDs protected with advanced ceramic or high-entropy coatings, together with encapsulants such as CYTOP or Parylene C, show drifts as low as ± 0.2 °C per 1,000 h and resist degradation up to ≈ 700 °C [106, 107]. For medium-demand scenarios, $Mo_{1-x}W_xS_2$ films and Cu–graphene/Al_2O_3 coatings offer strong stability and oxidation resistance, while standard platinum RTDs typically drift around ± 0.5 °C per year [108, 109]. In less critical uses, ferrous-alloy or copper RTDs with basic encapsulation are acceptable if periodic recalibration is planned; thin organic barriers such as SAALD-deposited Al_2O_3 can extend polymer-sensor lifetime in humid environments [110].

Thermal self-calibration and mechanical fatigue are controlled through material choice and device design. To keep self-heating below 0.05 °C, thin-film RTDs are driven with excitation currents of 1 mA or less; under these conditions, an internal thermal coefficient of 0.048 °C mW^{-1} has been reported, which limits drift to approximately 0.005 °C [111]. At intermediate performance levels, type-K thermocouples and nickel RTDs balance sensitivity and power dissipation, whereas for less demanding applications, NTC thermistors or wire-wound RTDs tolerate higher currents without sacrificing accuracy when calibration is maintained [112].

Regarding flexibility, sensors fabricated on PDMS/MXene substrates, CNT networks, or electronic textiles have withstood 10^5 or more bending cycles with minimal signal drift, while standard polymers and rigid ceramics are best reserved for fixed installations with negligible mechanical stress [113, 114].

In electromagnetically noisy environments, shielding with copper foils, Ag@GMs/PDMS composites, or other highly conductive coatings provides attenuation above 40 dB in the X-band and can exceed 100 dB with millimetric thicknesses [115]. For intermediate scenarios, NTC-RF filters and shielded FEP or PFA cables offer 20–50 dB of protection, which is suitable for general instrumentation, whereas clean laboratories or disposable devices may operate on unshielded substrates if the electromagnetic environment is benign [116].

As the focus shifts to humidity sensing, the design logic changes because transduction is governed by adsorption, diffusion, and dielectric variations. The primary targets are sensitivity, hysteresis, response and recovery times, temperature cross-sensitivity, and aging stability, followed by condensation tolerance, calibration drift, flexibility, and biocompatibility. Field performance is therefore determined by the deliberate choice of hydrophilic polymers, porous oxides, two-dimensional materials, and the encapsulation strategy.

For high-sensitivity use, hydrophilic, high–surface area networks make the decisive difference. Concentrated graphene oxide can deliver capacitance changes up to 37,800% across 15 to 95% RH with sub-second responses. Covalently functionalized $Ti_3C_2T_x$ MXene with polydopamine achieves sensitivities from 0.777 to 3.035, with response and recovery of 0.24 and 0.40 s across the full 0 to 100% RH range [117]. Graphene-oxide and cellulose-nanofiber laminates bend with curvatures from 0.012 to 0.260 cm^{-1} as RH rises from 25 to 85%, responding within 3 to 5 s [118]. MXene–nanocellulose resonant sensors reach sensitivities up to 20.54 Hz per %RH with hysteresis near 2.12% RH [119]. At intermediate performance levels, moderately thick PVA films show - 0.02167 MΩ per %RH with maximum hysteresis of 2.7% RH [120]. Porous SiO_2 on Si with 60 to 70% porosity and ≈ 15 nm pores yields resistance changes of about 10^4, with 3 and 9 s response and recovery [121]. Keratin–carbon-fiber composites report ≈ 633.12 pF per %RH with low hysteresis and fast dynamics. For basic or cost-driven devices, tissue paper shows ≈ 5.45 kΩ per %RH at low humidity between 58 and 75% RH and ≈ 411 Ω per %RH at high humidity between 75 and 100% RH [122], while non-crosslinked PEDOT:PSS remains functional with greater batch-to-batch variability [80, 123].

Hysteresis is curbed by combining network chemistry with barriers to liquid water. CYTOP encapsulation reduces hysteresis to negligible levels, especially with GOPS-crosslinked PEDOT:PSS, and maintains stable operation from 30 to 80% RH [80]. Covalently functionalized MXenes exhibit minimal hysteresis while operating across 0 to 100% RH [117]. Moderately crosslinked CNT-OH and graphene-oxide composites limit maximum hysteresis to about 3% RH over 30 to 95% RH [124]. Partially crosslinked PVA reaches 2.7% RH hysteresis over 12 to 98% RH [120], and CMC@graphene and GO/CNT-OH formulations further moderate hysteresis relative to untreated polymers [124, 125]. In contrast, non-crosslinked PVA, thick GO stacks, and standard polymer electrolytes show high hysteresis and are reserved for applications where precision is less critical, with standard PVA particularly prone to large hysteresis at high humidity due to capillary condensation in porous domains [120].

In practice, targets are set quantitatively for each variable by industry, and the sensor stack is assembled from material blocks that meet the required performance tier. Trade-offs are intrinsic: sensitivity against hysteresis, speed against aging, and cost against purity. These are managed by combining active layers, electrodes, substrates, and encapsulants with architectural controls on thickness, porosity, and effective area. Throughout, every decision remains anchored to the application constraints defined at the outset.

7. Printing techniques

Choosing the appropriate printing technique is crucial for fabricating temperature and humidity sensors, as each method provides specific benefits and constraints based on the intended application. Factors such as resolution, material compatibility, production scalability, and required sensor performance must be carefully evaluated. The following sections discuss critical parameters and operational considerations for several key printing methods, enabling informed decisions based on precise sensor requirements.

Screen printing is a contact technique in which a squeegee forces conductive ink through a patterned mesh onto the substrate, producing precise geometries for temperature and humidity sensors [126]. Transfer relies on the hydrodynamic pressure generated as the squeegee deflects the mesh toward the substrate and drives ink through the openings. Key equipment parameters include squeegee hardness, measured by Shore A durometer; angle, commonly between 60 and 75 degrees; and applied pressure, all of which govern deposition uniformity and final thickness [127]. Mesh characteristics strongly influence performance, particularly mesh count, such as 305/34 for high-precision temperature elements; wire diameter; mesh tension, ideally around 28–32 N cm^{-1}; and emulsion-over-mesh thickness, around 8–9 μm [6]. Ink rheology should be pseudoplastic and thixotropic, matched to the mesh so viscosity drops under shear for clean transfer and then recovers to prevent slumping. If viscosity is too low, the pattern bleeds and edges lose definition, whereas excessively high viscosity restricts ink passage through the mesh. For temperature sensors, precise control of squeegee speed, off-contact distance, and flood stroke pressure is required to achieve the target film thickness and curing profile that stabilizes the temperature coefficient of resistance [128, 129].

Inkjet printing is a non-contact digital technique in which piezoelectric or thermal actuators eject controlled droplets from micro-nozzles directly onto the substrate to form sensor patterns [130, 131]. The process hinges on accurate droplet formation, stable flight, and controlled impact, as droplet size and placement determine the final resolution and the electrical uniformity of the sensor [132]. Tight nozzle temperature control is essential for temperature-sensor fabrication because it sets ink viscosity and droplet consistency; reported optima range from 25 to 60 °C depending on the formulation, and small shifts can alter line width and continuity [133]. Viscosity must remain within approximately 0.001 to 1 Pa s to ensure reliable jetting. Silver nanoparticle inks often require tailored dispersants to preserve stability and prevent clogging during long print runs. Representative jetting windows include a driving voltage near 68 V with a 12 μs pulse, droplet spacing around 68 μm for continuous lines, and substrate heating selected to promote wetting without overspreading. Post-printing thermal treatments at 150 to 200 °C for 1 to 24 h are commonly used to sinter the film, lower resistivity, and achieve TCR values around $1.23–1.31 \times 10^{-3} K^{-1}$ for temperature sensing [134].

Aerosol jet printing is a non-contact direct-write method in which the ink is atomized into droplets of roughly 1 to 5 μm, passed through a virtual impactor, and aerodynamically focused by a nitrogen sheath through a ceramic nozzle before reaching the substrate [135]. The focused stream enables precise deposition at standoff distances of 1 to 11 mm and supports conformal printing on curved or textured surfaces, which is valuable for integrated temperature and humidity systems. Atomizer flow, typically 800 to 1500 sccm, regulates aerosol generation and governs printed line width and thickness. Sheath flow, often 40 to 90 sccm, sets the focusing strength and permits features near 10 μm. The viscosity window is broad, about 0.001 to 1 Pa s, which accommodates higher solids and helps raise conductivity. Substrate temperatures near 40 to 60 °C curb excessive solvent loss, while dispensing speeds of 1 to 5 mm s^{-1} support smooth lines. Controlled atmospheres further protect sensitive materials [136]. Post-processing thermal anneals between 150 and 350 °C activate nanoparticle sintering and can deliver TCR values approaching 3.7×10^{-3} °C^{-1} [81, 137].

Roll-to-roll gravure uses a continuously fed web. Ink fills the engraved cells of a rotating gravure cylinder, a doctor blade removes excess ink, and the patterned ink transfers to the moving substrate under regulated pressure and speed [138]. The method supports high-throughput fabrication of temperature sensors on flexible substrates at line speeds up to 6 m min^{-1}, with registration accuracy within ± 20 μm [59, 78].

Doctor-blade material, commonly steel, together with a blade angle near 60 degrees and carefully set pressure, ensures effective wiping while retaining the correct ink volume in the cells. Ink viscosity, typically 10 to 100 cP, works in tandem with press speed to set the capillary number, defined as $Ca = \mu U/\gamma$, which governs cell filling, wiping quality, and transfer. Lower capillary numbers improve pattern fidelity but limit the maximum speed [139].

Flexography meters ink with an anilox roll, whose engraved cells deliver a controlled volume to a flexible relief plate, which then transfers the image to the substrate. Proper anilox selection is critical, with typical cell volumes of 1.5 to 8.0 BCM at 250 to 700 LPI, and inks are kept within about 600 to 2000 cP to avoid starvation or flooding [140]. Post-processing for printed temperature sensors often includes thermal annealing between 150 and 600 °C to enhance crystallinity and electrical performance, multilayer encapsulation such as Parylene C combined with Al_2O_3 to improve barrier properties, and electrical characterization by impedance spectroscopy from 10 kHz to 5 MHz to verify functionality and calibration accuracy [141, 142].

Stereolithography and fused deposition modeling (FDM) are less common choices for temperature and humidity sensing, yet they offer useful routes when geometry, packaging, or rapid iteration are paramount. Stereolithography cures photo-resins into high-resolution structures that can later be metallized or coated with hygroscopic layers, enabling capacitive elements or packages that host the sensing interface. It has been used to build microscale housings and flow channels that integrate printed electrodes or films for environmental monitoring [143, 144]. Fused deposition modeling extrudes thermoplastic composites that carry conductive fillers, allowing resistive temperature elements to be printed directly into parts. Demonstrations with graphene–PLA and other carbon-loaded filaments have delivered workable thermistors for real-time temperature readout on additively manufactured platforms [145, 146]. Although rarer, FDM has also produced interdigitated structures for capacitive humidity sensing, where the printed geometry acts as the electrode frame and a post-applied hygroscopic layer provides transduction [147].

In **Figure 5**, the end-to-end workflow for printing temperature and humidity sensors begins with digital design preparation in Step 1 and concludes with final inspection. The process starts by defining the sensor layout in CAD, including conductor traces, sensing elements, and contact pads sized to the required measurement range and resolution. In Step 2, the pattern is transferred to the selected substrate using screen printing, inkjet printing, AJP, roll-to-roll gravure, or flexography, chosen according to feature size, throughput, and material compatibility. Process windows are then set by tuning ink rheology, squeegee speed or nozzle actuation, mesh tension or cell geometry, and web speed so that layer thickness and pattern fidelity remain uniform across the panel or web.

After deposition, Step 3 converts wet films into functional layers through drying and curing in convection or infrared ovens, or by microwave and RF drying. Nanoparticle inks may be consolidated by photonic flash sintering or UV curing to activate conduction pathways and stabilize properties [148, 149]. In Step 4, encapsulation adds barrier coatings against moisture and contaminants, as well as biocompatible layers when required for on-skin or plant-interface use. The workflow ends with optical inspection to verify pattern integrity, followed by electrical tests that confirm resistance, temperature coefficient of resistance, and humidity

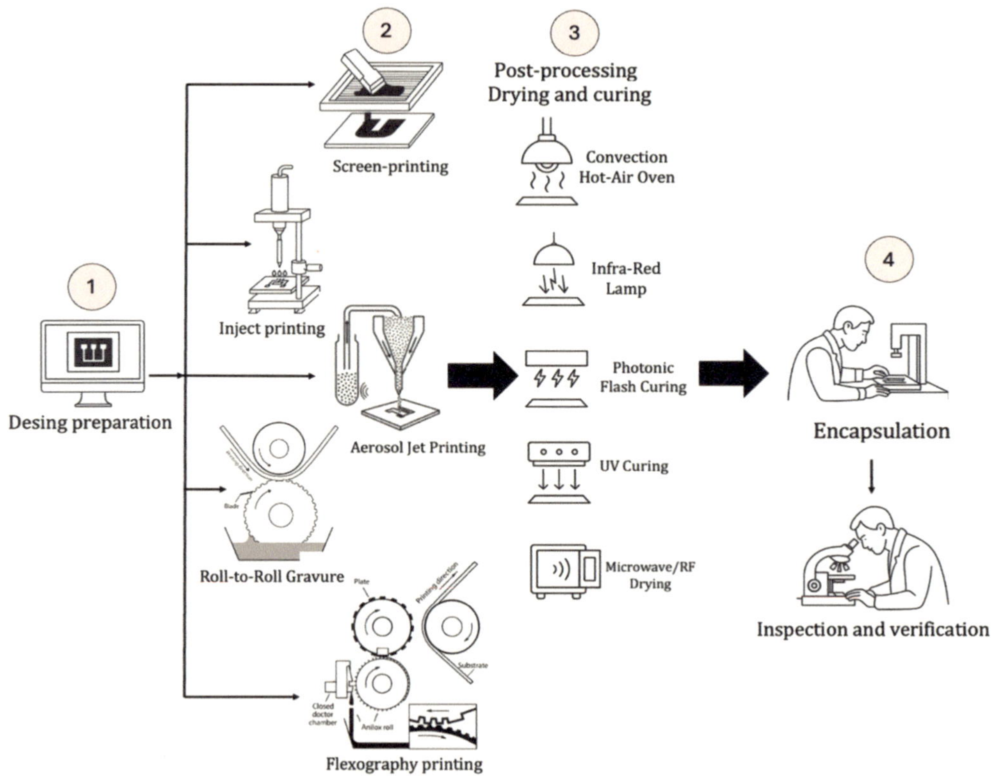

Figure 5.
Diagram showing the manufacturing process flow for printed temperature and humidity sensors, starting with design preparation followed by various printing methods including screen printing, inkjet, aerosol jet, roll-to-roll gravure, and flexography. The process continues with drying and curing methods such as convection, infrared, photonic flash, UV, and microwave drying, ending with encapsulation and inspection.

response meet specifications. Together, these steps define a robust and scalable route for manufacturing high-performance printed sensors.

Figure 5 distills how fabrication choices shape performance in low-cost printed temperature and humidity sensors. In temperature sensing, resolution and film thickness govern thermal mass and TCR linearity. In humidity sensing, the microstructure and uniformity of thin films control sensitivity, hysteresis, and response time.

Inkjet provides the finest features, which benefit interdigital electrodes and narrow meander RTDs. Screen printing, with moderate resolution, excels when uniform, well-controlled thickness and low unit cost at volume are paramount. DIW and FDM favor thick or structural features rather than fine electrodes, while stereolithography contributes precise geometries for substrates and packages. In material compatibility, screen printing and DIW accommodate the broadest range of metallic, ceramic, polymeric, and composite inks, whereas inkjet is best for low-viscosity dispersions such as PEDOT:PSS, GO, or MXene. Throughput separates the methods as well: screen printing scales cleanly to roll-to-roll for large areas, inkjet balances agility with medium throughput, and DIW operates more slowly but enables thick films.

Taken together, **Figure 6** serves as a design guide. The appropriate process depends on target metrics, budget, available materials, and the tools at hand. Researchers should select the route that aligns most directly with their performance needs and practical constraints.

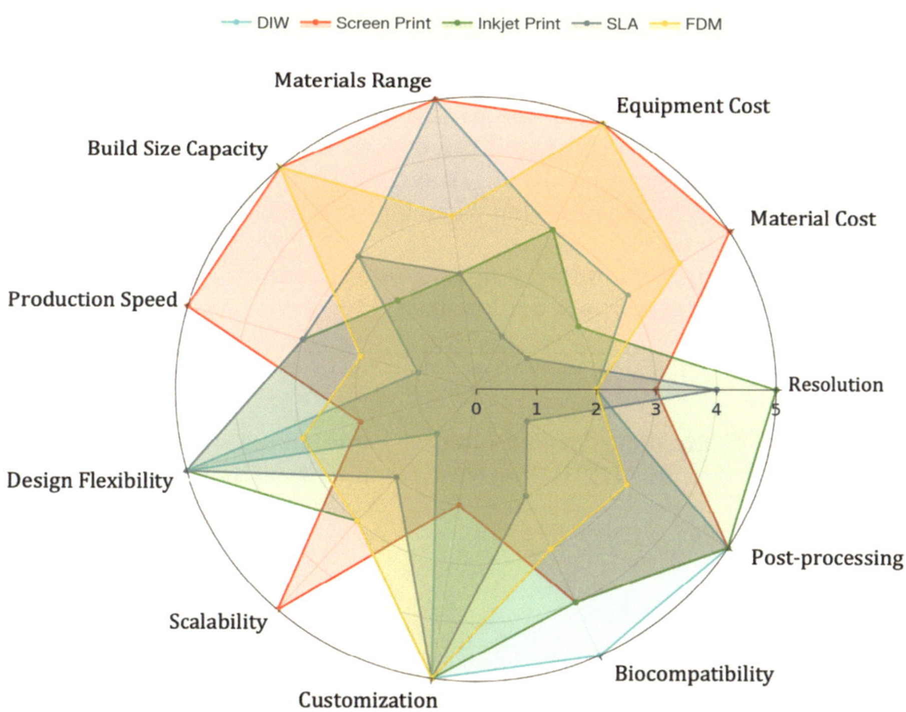

Figure 6.
Radar chart comparing the performance of DIW, screen printing, inkjet printing, SLA, and FDM technologies across key manufacturing metrics including resolution, post-processing, biocompatibility, customization, scalability, design flexibility, production speed, and material and equipment costs.

8. Data processing and low-power AI for printed flexible temperature and humidity sensors

Printed thin-film sensors produce dense data streams, whose reliability is limited by stochastic noise, slow drift, and thermo-hygrometric cross-effects. Throughout this section, the sensing channel is described by the composite model [150].

$$v(t) = G(T(t), H(t)) + n(t) + d(t) \tag{1}$$

Where $G(\cdot)$ is the ideal transfer function from temperature T and relative humidity, $H, n(t)$ denotes conducted and radiated Gaussian noise with variance σ^2; and $d(t)$ captures quasi-static drift such as polymer aging and electrode creep [151], represented compactly by

$$d(t) = \sum_{k=0}^{M} a_k t^k \tag{2}$$

with coefficients $\{a_k\}$ estimated per device, in multi-track arrays, temperature–humidity cross-interference becomes prominent during fast thermal transients; compensation, therefore, includes a temperature-rate term while keeping the structure of Equation 1 intact.

In practice, the pipeline is instantiated on the devices introduced earlier in the chapter: screen-printed silver RTDs, inkjet-printed silicon-carbide (SiC) thermistors, and polymer-composite capacitive humidity sensors on flexible substrates. Calibration uses the same film-specific models already presented for those materials so that the per-device coefficients referenced around (3)–(5) reflect thickness control, microstructure, and curing/annealing achieved by screen and inkjet printing. This preserves traceability from fabrication parameters such as mesh count, emulsion thickness, drop spacing, solvent system, and thermal profile to inference accuracy, avoiding re-fitting when a print run changes only process variables.

Signal conditioning maximizes the signal-to-noise ratio prior to quantization. The analog front-end applies a first-order low-pass filter with a cutoff

$$f_c = \frac{1}{2\pi R_f C_f} \tag{3}$$

where R_f and C_f are the front-end filter elements, so out-of-band interference is attenuated before sampling [152]. Discrete-time smoothing then uses a finite impulse response (FIR),

$$y[n] = \sum_{k=0}^{N-1} h[k]\, x[n-k] \tag{4}$$

where $x[n]$ is the sampled sensor output, $h[k]$ the impulse response, and $y[n]$ the filtered sequence; N and $\{h[k]\}$ are selected to satisfy latency and pass-band ripple constraints [153]. Samples are standardized online via

$$x' = \frac{x - \mu}{\sigma} \tag{5}$$

where μ and σ are the running mean and standard deviation computed over a sliding window of multiple samples to track slow environmental shifts without blurring dynamics [154]. All conditioning arithmetic is executed in fixed-point to respect milliwatt-class energy budgets, and per-device calibration constants are stored in non-volatile memory for deterministic startup and lot-to-lot reproducibility.

Baseline drift is removed by estimating and subtracting a smooth background aligned with the most-cited Whittaker/ALS formulations. Specifically, the baseline sequence $b = \{b_i\}_{i=1}^{N}$ is obtained by solving.

$$\min_{b} \sum_{i=1}^{N} w_i\,(s_i - b_i)^2 + \lambda \sum_{i=3}^{N} \left(\Delta^2 b_i\right)^2, \quad \Delta^2 b_i = b_i - 2b_{i-1} + b_{i-2} \tag{6}$$

where $s = \{s_i\}$ are the observations, w_i are adaptive/asymmetric weights that prevent peaks from leaking into the baseline, and λ tunes smoothness via the second-difference operator Δ^2. Because polymeric dielectrics and printed conductors evolve under humidity and temperature cycling, the second-difference baseline in Equation 6 is paired with scheduled background recalibration [155]. Its cadence reflects the durability tests reported in the chapter, including thermal cycling, high-RH soaks, and bend fatigue, ensuring that drift captured during qualification becomes the prior for field operation. Hysteresis in humidity films is treated as a materials-driven phenomenon and encoded as a rate-aware correction layered before probabilistic tracking, ensuring that mechanical relaxation and sorption kinetics do not contaminate state estimates.

After detrending, a compact auto-calibration stage corrects residual gain and offset using a tiny regression model trained with

$$L = |y - \hat{y}|_2^2 \tag{7}$$

where y is the calibrated target and $\hat{y}$ the model output. The trained model is quantized and deployed to keep latency and flash footprint within embedded limits [156].

Fast environmental fluctuations are tracked with a discrete Kalman filter using the standard linear state-space relations. The prediction step is

$$\hat{x}_{k|k-1} = F_k \hat{x}_{k-1|k-1} + B_k u_k, P_{k|k-1} = F_k P_{k-1|k-1} F_k^{\top} + Q_k \tag{8}$$

and the update step is

$$K_k = P_{k|k-1} H_k^{\top} \left(H_k P_{k|k-1} H_k^{\top} + R_k\right)^{-1}, \; \hat{x}_{k|k} = \hat{x}_{k|k-1} + K_k\left(z_k - H_k \hat{x}_{k|k-1}\right) \tag{9}$$

where $\hat{x}_{k|k}$ is the posterior state estimate, P its covariance, F_k the state transition, $B_k u_k$ is a known input, Q_k the process noise covariance, H_k the measurement model, R_k the measurement noise covariance, z_k the observation, and K_k the Kalman gain. With F_k and H_k set to identity for random-walk tracking of slowly varying baselines, the implementation fits comfortably on low-power MCUs while providing uncertainty metrics used later for confidence-weighted telemetry [157, 158].

Low-power inference follows TinyML practices, which refer to design strategies that allow machine learning models to run on highly resource-constrained devices like microcontrollers [159, 160]. These strategies focus on minimizing memory, computation, and energy use. In this work, we apply uniform post-training quantization to map model parameters to low-bit integers, enabling efficient integer multiply–accumulate operations directly on embedded processors while maintaining accuracy.

$$q = \mathrm{round}\left(\frac{x}{\Delta}\right), \tag{10}$$

with quantization step

$$\Delta = \frac{x_{max} - x_{min}}{2^b - 1}, \tag{11}$$

where b is the bit-width, and x_{min}, x_{max} define the calibration range [161]. For classification and regression tasks common to temperature/humidity monitoring, MicroNet-class models with only a few thousand parameters provide sufficient temporal context when combined with the conditioning stack above. Energy is budgeted using the CMOS switching relation

$$E_{\mathrm{inf}} = C\,V^2\,N_{\mathrm{ops}}, \tag{12}$$

where C is the effective switched capacitance, V the supply voltage, and N_{ops} the operation count. We use Equation 12 as a design driver: filter order, feature dimensionality, and model depth are selected to minimize joules per decision at the required accuracy. Early-exit mechanisms terminate inference once confidence is high, shrinking N_{ops} without sacrificing detection performance [162, 163].

Anomaly detection and communication are co-designed to minimize radio energy. A lean feature vector, comprising incremental statistics such as running mean and variance, dominant spectral peaks, and change-point indicators, feeds a depth-limited isolation forest executed locally; only a compact anomaly index is transmitted when alarms are confirmed [164]. This shifts energy from radio to on-device computation, where it is more economical and aligns with the sustainability goals of printed, battery-constrained nodes. The event-only telemetry integrates naturally with the uncertainty provided by the Kalman filter in Equations 8 and 9, allowing the node to flag degraded regimes such as condensation or fouling rather than silently biasing outputs.

Implementation adheres to four practical guidelines that respect the constraints of flexible, printed platforms. First, aggressive weight compression and low-bit quantization preserve accuracy while fitting the complete stack, including calibration constants, filters, and the model, within tens of kilobytes of flash. Second, batch scheduling groups ADC reads in short bursts and returns the core to sleep for most of the duty cycle without violating latency bounds set by the analog front end in Equation 3. Third, sensor-centric coprocessing offloads simple convolutions or thresholding to in-sensor processing units when available, waking the MCU only on events. Fourth, in-situ updates adjust a tiny subset of parameters under strict bounds, preventing instability while accommodating slow device aging observed in durability tests.

Energy management is reported alongside accuracy so that trade-offs are explicit. Using Equation 12, the choice of filter order, feature set, and model size is justified by total joules per decision, not only by error metrics. Because radios dominate power in many flexible nodes, edge summarization and event-only telemetry are treated as integral parts of the model design, with communication schedules co-optimized against inference cadence to meet milliwatt-class budgets.

Validation follows the test philosophy used for the printed devices in this chapter. Each unit undergoes per-device calibration at first power-up; cross-sensitivity checks in two-factor chamber sweeps of temperature and humidity; latency measurements at the sampling rates dictated by the analog front end; and stability assessment under the same durability protocols used for materials screening. Reported results include accuracy with confidence bounds derived from the Kalman covariance in Equations 8 and 9, energy per inference computed with Equation 12, and end-to-end power, including radio duty cycling, making figures of merit reproducible across laboratories and traceable to fabrication parameters.

Finally, the stack built around Equations 1 through 12 maps cleanly onto the three deployment modes highlighted in this book. For continuous wearable monitoring, strict energy caps dictate aggressive duty cycling and the smallest viable model; for environmental logging, adaptive sampling schedules dominate, with longer sleep intervals and event-only uploads; for industrial control, latency constraints prioritize filter settings and deterministic inference timing. Crucially, the same printed hardware, including screen-printed RTDs, inkjet-printed SiC thermistors, and polymer-composite humidity elements, serves all three cases by changing only the scheduling policy and the quantized model. Fabrication, calibration, and reliability practices remain unchanged, ensuring that the data-processing and AI layer is a natural extension of the materials and process engineering emphasized throughout the chapter.

9. Conclusions

Printed flexible temperature and humidity sensors have demonstrated significant potential for scalable, low-cost, and conformable sensing solutions, driven by multiple printing methods. Systematic analysis and comparative evaluation of materials such as metal nanoparticles, conductive polymers, nanocomposites, and two-dimensional materials have clarified how ink formulation, curing parameters, and pattern resolution directly influence critical sensor characteristics like sensitivity, stability, response time, and mechanical robustness. A particularly promising contribution highlighted is the integration of AI-driven calibration and drift-correction strategies, enabling printed sensors to maintain accuracy despite environmental variations and extend their operational lifespan in practical settings.

However, the field still faces important limitations and standardization challenges. Consistent reporting practices for comprehensive sensor performance, including energy consumption, flexibility under repeated mechanical deformation, and reliable encapsulation methods, remain critical for broader adoption. Additionally, harmonized calibration procedures and standardized environmental testing protocols are essential to facilitate meaningful comparisons and accelerate technology transfer from laboratory-scale prototypes to industry-ready products.

Moving forward, research should prioritize the development of ambient-processed inks suitable for high-throughput roll-to-roll production, combined sensor-power-communication platforms, and comprehensive benchmarking frameworks that enable direct comparative assessment. Addressing these areas will be key to transitioning printed flexible sensors from proof-of-concept devices toward robust, standardized, and commercially viable sensor systems for widespread deployment.

Acronyms and abbreviations

AI	Artificial intelligence
AJP	Aerosol jet printing
Al_2O_3	Aluminum oxide
BP	Black phosphorus
CAB	Cellulose acetate butyrate
CNT	Carbon nanotube
COC	Cyclic olefin copolymer
CS	Chitosan
CYTOP	Cyclic transparent optical polymer
DIW	Direct ink writing
EMF	Electromotive force
EtC	Ethyl cellulose
FDM	Fused deposition modeling
FET	Field effect transistor
GO	Graphene oxide
GOPS	(3-glycidyloxypropyl)trimethoxysilane
IDC	Interdigital capacitor
IDE	Interdigitated electrode
IDTechEx	IDTechEx (market research company)
IEC	International Electrotechnical Commission
In_2O_3	Indium oxide
IoT	Internet of Things
ITO	Indium tin oxide
J-type	J-type thermocouple
K-type	K-type thermocouple
La-doped $BaTiO_3$	Lanthanum-doped barium titanate
LCP	Liquid crystal polymer
ML	Machine learning
MoS_2	Molybdenum disulfide
MXene	Transition metal carbide/nitride two-dimensional material
NFC	Near-field communication
NTC	Negative temperature coefficient
PCL	Polycaprolactone
PDMS	Polydimethylsiloxane
PEDOT:PSS	Poly(3,4 ethylenedioxythiophene):polystyrene sulfonate
PET	Polyethylene terephthalate
PI	Polyimide
PN	Polymeric naphthalate
PVA	Polyvinyl alcohol
PEN	Polyethylene naphthalate
Pt	Platinum
PVAC	Polyvinyl acetate
PTC	Positive temperature coefficient
QCM	Quartz crystal microbalance
R2R	Roll-to-roll

DOI: http://dx.doi.org/10.5772/intechopen.1012577

RH	Relative humidity
r-GO	Reduced graphene oxide
RTD	Resistance temperature detector
SA	Sodium alginate
SiC	Silicon carbide
SiCN	Silicon carbonitride
SWCNT	Single-walled carbon nanotube
TCR	Temperature coefficient of resistance
UV	Ultraviolet
WO_3	Tungsten trioxide
ZnO	Zinc oxide

Nomenclature

Symbol	Description	Unit	Symbol	Description	Unit
A	Effective electrode area	m^2	R	Electrical resistance	Ω
B	Thermistor characteristic parameter	K	R0	Baseline resistance	Ω
C	Capacitance	F	R1	Measured resistance	Ω
C0	Baseline capacitance	F	RH	Relative humidity	%
C1	Measured capacitance	F	S	Sensitivity	various
Cf	Filter capacitance	F	T	Temperature	°C
D	Dielectric layer thickness or electrode separation	M	TCR	Temperature coefficient of resistance	$°C^{-1}$
F	Frequency	Hz	V	Voltage	V
G	Transfer function	–	V0	Reference voltage	V
H	Relative humidity	%	ΔR	Change in resistance	Ω
I	Electric current	A	ΔC	Change in capacitance	F
L	Sensor length	M	λ	Smoothing parameter	–
N	Drift/noise (general)	–	σ	Standard deviation	–
P	Power	W	μ	Mean	–
Q	Process noise covariance	–	f_c	Signal-to-noise ratio	dB

Author details

Juan C. Rubio and Martin Bolduc*
Department of Mechanical Engineering, Université du Québec à Trois-Rivières, Trois-Rivières, Quebec, Canada

*Address all correspondence to: martin.bolduc2@uqtr.ca

References

[1] Barmpakos D, Kaltsas G. A review on humidity, temperature and strain printed sensors—current trends and future perspectives. Sensors. 2021; **21**(3):739. DOI: 10.3390/s21030739

[2] da Costa TP, *et al.* A systematic review of real-time monitoring technologies and its potential application to reduce food loss and waste: Key elements of food supply chains and IoT technologies. Sustainability. 2022;**15**(1):614. DOI: 10.3390/SU15010614.

[3] Rao X, *et al.* Review of optical humidity sensors. Sensors. 2021;**21** (23):8049. DOI: 10.3390/S21238049.

[4] Kuzubasoglu BA. Recent studies on the humidity sensor: A mini review. ACS Applied Electronic Materials. 2022;**4**(10):4797–4807. DOI: 10.1021/ACSAELM.2C00721

[5] Tobjörk D, Österbacka R. Paper electronics. Advanced Materials. 2011;**23**(17):1935–1961. DOI: 10.1002/ADMA.201004692

[6] Rubio JC, Bolduc M. Screen printing for energy storage and functional electronics: A review. Electronic Materials. 2025;**6**(2):7. DOI: 10.3390/ELECTRONICMAT6020007

[7] Kanon K, Sharif SS, Irfan A, Sharif A. Inorganic film materials for flexible electronics: A brief overview, properties, and applications. Engineering Reports. 2024;**6**(11): e13006. DOI: 10.1002/ENG2.13006

[8] He Y, Han Z, Ren J, Hou H, Pan D. An overview of flexible sensing nanocomposites. Advanced Composites and Hybrid Materials. 2024;**7**(5):1–15. DOI: 10.1007/S42114-024-00954-Y

[9] Liu E, Cai Z, Ye Y, Zhou M, Liao H, Yi Y. An overview of flexible sensors: Development, application, and challenges. Sensors (Basel). 2023;**23** (2):817. DOI: 10.3390/S23020817

[10] 'Printed and flexible sensors 2020-2030: Technologies, players, forecasts: IDTechEx'. [Online]. Available: https://www.idtechex.com/en/research-report/printed-and-flexible-sensors-2020-2030-technologies-players-forecasts/755. [Accessed: 2025-June-8]

[11] Claggett TJ, Worrall RW, Clayton WA, Lipták BG. Resistance Temperature Detectors (RTDs). Temperature Measurement. 2022;75–84. DOI:10.1201/9781003063919-13

[12] Huebener RP. Effect of phonon drag on the electrical resistivity of metals. Physical Review. 1966;**146** (2):502. DOI: 10.1103/PhysRev.146.502

[13] Dames C. Resistance temperature detectors. Encyclopedia of Microfluidics and Nanofluidics. 2008;1782–1790. DOI:10.1007/978-0-387-48998-8_1354

[14] Fischer B, Behrends A, Freund D, Lupton DF, Merker J. High temperature mechanical properties of the platinum group metals. Platinum Metals Review. 1999;**43**(1):18–28. DOI: 10.1595/003214099X4311828

[15] Goupil FL, Payrot G, Khiev S, Smaal W, Hadziioannou G. Fully printed sensors for in situ temperature, heat flow, and thermal conductivity measurements in flexible devices. ACS

Omega. 2023;**8**(9):8481–8487. DOI: 10.1021/acsomega.2c07590

[16] Jäger J, Schwenck A, Walter D, Bülau A, Gläser K, Zimmermann A. Inkjet-printed temperature sensors characterized according to standards. Sensors. 2022;**22**(21):8145. DOI: 10.3390/s22218145

[17] Jacovelli PB, Zinke OH. The thermocouple revisited: The benedicks and seebeck effects. Journal of Non-Equilibrium Thermodynamics. 2019;**44**(4):373–383. DOI: 10.1515/JNET-2018-0063

[18] Bajzek TJ. Thermocouples: A sensor for measuring temperature. IEEE Instrumentation & Measurement Magazine. 2005;**8**(1):35–40. DOI: 10.1109/MIM.2005.1405922

[19] Gierczak M, Prazmowska-Czajka J, Dziedzic A. Thermoelectric mixed thick-/thin Film microgenerators based on constantan/silver. Materials. 2018;**11**(1):115. DOI: 10.3390/MA11010115

[20] Offenzeller C, Knoll M, Jakoby B, Hilbers W. Screen-printed, pure carbon-black thermocouple fabrication and seebeck coefficients. Sensors (Basel). 2019;**19**(2):403. DOI: 10.3390/S19020403

[21] Talplacido NC, Cumming DJ. Printed carbon black thermocouple as an in situ thermal sensor for lithium-Ion cell. Batteries. 2024;**10**(3):78. DOI: 10.3390/BATTERIES10030078

[22] Bodić MZ, Aleksić SO, Rajs VM, Damnjanović MS, Kisić MG. Thermally coupled NTC chip thermistors: Their properties and applications. Sensors. 2024;**24**(11):3547. DOI: 10.3390/S24113547

[23] Yan C, Wang J, Lee PS. Stretchable graphene thermistor with tunable thermal index. ACS Nano. 2015;**9**(2):2130–2137. DOI: 10.1021/NN507441C

[24] Turkani VS, Maddipatla D, Narakathu BB, Bazuin BJ, Atashbar MZ. A carbon nanotube based NTC thermistor using additive print manufacturing processes. Sens Actuators A Physics. 2018;**279**:1–9. DOI: 10.1016/J.SNA.2018.05.042.

[25] Shi L, Zhang N, Li L, Ding S, Li W, Li K. Printed flexible temperature sensors: Principles, materials, processes, and applications. Applied Materials today. 2025;**43**. 102665. DOI: 10.1016/J.APMT.2025.102665.

[26] Sekiguchi T, Sotoma S, Harada Y. Fluorescent nanodiamonds as a robust temperature sensor inside a single cell. Biophysics and Physicobiology. 2018;**15**:229. DOI: 10.2142/BIOPHYSICO.15.0_229.

[27] Chowdhury HR, Han M. Fiber optic temperature sensor system using air-filled fabry–pérot cavity with variable pressure. Sensors (Basel). 2023;**23**(6):3302. DOI: 10.3390/S23063302

[28] Koegl M, Delwig M, Zigan L. Characterization of fluorescence tracers for thermometry and film thickness measurements in liquid coolants relevant for thermal management of electric and electronic components. Sensors (Basel). 2022;**22**(22):8892. DOI: 10.3390/S22228892

[29] Xue K, *et al.* A sensitive and reliable organic fluorescent nanothermometer for noninvasive temperature sensing. Journal of the American Chemical Society. 2021;**143**(35):14147–14157. DOI: 10.1021/JACS.1C04597.

[30] Meitei SN. Partial discharge detection using Piezoelectric sensors on power transformer: A review. IEEE Sensors Journal. 2024;**24**(9):13730–13742. DOI: 10.1109/JSEN.2024.3379037

[31] Gautschi G. Piezoelectric Sensors. Piezoelectric Sensorics. 2002;73–91. DOI:10.1007/978-3-662-04732-3_5

[32] Ghemari Z, Belkhiri S, Saad S. A piezoelectric sensor with high accuracy and reduced measurement error. Journal of Computational Electronics. 2024;**23**(2):448–455. DOI: 10.1007/S10825-024-02134-Z

[33] Li J, *et al.* 3D printing of flexible piezoelectric composite with integrated sensing and actuation applications. Advanced Powder Materials. 2024;**3**(5):100226. DOI: 10.1016/J.APMATE.2024.100226

[34] Zhou X, *et al.* Piezoelectric-pneumatic material jetting printing for non-contact conformal fabrication of high-temperature thick-film sensors. Additive Manufacturing. 2024;**83**:104058. DOI: 10.1016/J.ADDMA.2024.104058.

[35] Zhang Z, *et al.* Printed flexible capacitive humidity sensors for field application. Sensors and Actuators, B: Chemical. 2022;**359**:131620. DOI: 10.1016/J.SNB.2022.131620.

[36] Farooq Z, *et al.* Investigation of relative humidity-sensing performance of capacitive and resistive type sensor based on TDTBPPNi metalloporphyrin dielectric layer. Bulletin of Materials Science. 2021;**44**(2):1–10. DOI: 10.1007/s12034-021-02443-0.

[37] Khan MU, *et al.* Wide range and highly linear signal processed systematic humidity sensor array using Methylene Blue and Graphene composite. Scientific Reports. 2021;**11**(1):1–10. DOI: 10.1038/S41598-021-95977-6.

[38] Hassan G, Sajid M, Choi C. highly sensitive and full range detectable humidity sensor using PEDOT:PSS, methyl red and graphene oxide materials. Scientific Reports. 2019;**9**(1):1–10. DOI: 10.1038/S41598-019-51712-W

[39] Li S, *et al.* Humidity-sensitive chemoelectric flexible sensors based on metal-air redox reaction for health management. Nature Communications. 2022;**13**(1):1–10. DOI: 10.1038/s41467-022-33133-y.

[40] Wang X, Deng Y, Chen X, Jiang P, Cheung YK, Yu H. An ultrafast-response and flexible humidity sensor for human respiration monitoring and noncontact safety warning. Microsystems & Nanoengineering. 2021;**7**(1):1–11. DOI: 10.1038/S41378-021-00324-4

[41] McGhee JR, Sagu JS, Southee DJ, Evans PSA, Wijayantha KGU. Printed, fully metal oxide, capacitive humidity sensors using conductive indium tin oxide inks. ACS Applied Electronic Materials. 2020;**2**(11):3593–3600. DOI: 10.1021/acsaelm.0c00660

[42] Farahani H, Wagiran R, Hamidon MN. Humidity sensors principle, mechanism, and fabrication technologies: A comprehensive review. Sensors (Basel, Switzerland). 2014;**14**(5):7881. DOI: 10.3390/S140507881

[43] Estefani RB, Cárdenas AAB, Usman IB, de Abreu GJP, Serbena JPM. 'Relative humidity detection in TiO 2 quantum dots and multi-walled carbon nanotubes composite sensors through

electrical impedance spectroscopy'. ACS Applied Electronic Materials. 2025;**7**(10):4413–4424. DOI: 10.1021/ACSAELM.5C00130

[44] Nenova Z, Nenov T, Kozhukharov S, Machkova M. Humidity sensing elements based on Zn-doped TiO2 films prepared via a Sol-gel Method. Materials Science. 2015;**21**(2):191–197. DOI: 10.5755/J01.MS.21.2.6244

[45] Anisimov YA, Evitts RW, Cree DE, Wilson LD. Polyaniline/Biopolymer composite systems for humidity sensor applications: A review. Polymers (Basel). 2021;**13**(16):2722. DOI: 10.3390/POLYM13162722

[46] Dai J, *et al.* Ultrafast Response Polyelectrolyte Humidity Sensor for Respiration Monitoring. ACS Applied Materials & Interfaces. 2019;**11**(6):6483–6490. DOI: 10.1021/ACSAMI.8B18904.

[47] Kim S, Han SG, Koh YG, Lee H, Lee W. Colorimetric humidity sensor using Inverse Opal photonic gel in hydrophilic ionic liquid. Sensors. 2018;**18**(5):1357. DOI: 10.3390/S18051357

[48] Torabi-Monfared H, Sherafat L, Doroodmand MM, Eshghi F. Fluorescent lamp tungsten filament thermionic emission gun as a novel humidity optical sensor. Scientific Reports. 2021;**11**(1):1–12. DOI: 10.1038/S41598-021-97688-4

[49] Lee H, *et al.* Optical humidity sensors based on lead-free Cu-based perovskite nanomaterials. Nanoscale Advances. 2022;**4**(16):3309–3317. DOI: 10.1039/D2NA00168C.

[50] Emperhoff S, Eberl M, Dwertmann T, Wöllenstein J. On the influence of humidity on a thermal conductivity sensor for the detection of hydrogen. Sensors (Basel). 2024;**24**(9):2697. DOI: 10.3390/S24092697

[51] Emperhoff S, Eberl M, Dwertmann T, Wöllenstein J. Humidity impact on thermal conductivity sensors. Proceedings. 2024;**97**(1):93. DOI: 10.3390/PROCEEDINGS2024097093

[52] Lee CY, Lee GB. Humidity sensors: A Review. Sensor Letters. 2005;**3**(1):1–15. DOI: 10.1166/SL.2005.001

[53] Yi R, Peng B, Zhao Y, Nie D, Chen L, Zhang L. Quartz crystal microbalance humidity sensors based on structured graphene oxide membranes with magnesium Ions: Design, mechanism and performance. Membranes (Basel). 2022;**12**(2):125. DOI: 10.3390/MEMBRANES12020125

[54] Chen Q, *et al.* Humidity-sensing properties of a BiOCl-coated quartz crystal microbalance. ACS Omega. 2020;**5**(30):18818–18825. DOI: 10.1021/ACSOMEGA.0C01946/ASSET/IMAGES/LARGE/AO0C01946_0008.JPEG.

[55] Zheng Z, Zhang G, Wang X, Kong X. Comparative study of gravimetric humidity sensor platforms based on CMUT and QCM. Micromachines (Basel). 2022;**13**(10):1651. DOI: 10.3390/MI13101651

[56] Niro G, Marasco I, Rizzi F, D'Orazio A, Grande M, De Vittorio M. Design and fabrication of a flexible gravimetric sensor based on a thin-film bulk acoustic wave resonator. Sensors (Basel). 2023;**23**(3):1655. DOI: 10.3390/S23031655

[57] Soukup R, Hamáček A, Řeboun J, 'Advanced screen printing for the fabrication of organic humidity

sensors', *2012 4th Electronic System-Integration Technology Conference, ESTC 2012*, 2012, DOI: 10.1109/ESTC.2012.6542183.

[58] Kutzner C, Lucklum R, Torah R, Beeby S, Tudor J, Novel Screen Printed Humidity Sensor on Textiles for Smart Textile Applications.

[59] Kang H, *et al.* Fully roll-to-roll gravure printable wireless (13.56 MHz) sensor-signage tags for smart packaging. Scientific Reports. 2014;**4**. DOI: 10.1038/srep05387.

[60] He P, *et al.* Fully printed high performance humidity sensors based on two-dimensional materials. Nanoscale. 2018;**10**(12):5599–5606. DOI: 10.1039/c7nr08115d.

[61] Wu M, Wu Z, Jin X, Lee JH.A highly sensitive FET-type humidity sensor with inkjet-printed Pt-In2O3 nanoparticles at room temperature. Nanoscale Research Letters. 2020;**15**(1). DOI: 10.1186/s11671-020-03426-6

[62] Pereira NM, *et al.* Aerosol-printed MoS2Ink as a high sensitivity humidity sensor. ACS Omega. 2022;**7**(11):9388–9396. DOI: 10.1021/acsomega.1c06525.

[63] Lim WY, Goh CH, Yap KZ, Ramakrishnan N.One-step fabrication of paper-based inkjet-printed graphene for breath monitor sensors. Biosensors (Basel). 2023;**13**(2). DOI: 10.3390/bios13020209

[64] Beniwal A, Ganguly P, Aliyana AK, Khandelwal G, Dahiya R. Screen-printed graphene-carbon ink based disposable humidity sensor with wireless communication. Sensors and Actuators, B: Chemical. 2023;**374**. DOI: 10.1016/j.snb.2022.132731.

[65] Guo P, *et al.* An all-printed, fast-response flexible humidity sensor based on hexagonal-WO3 nanowires for multifunctional applications. Advanced Materials. 2023;**35**(41). DOI: 10.1002/adma.202304420

[66] Jiang L, *et al.* 'High-sensitivity fully printed flexible BaTiO 3 - Based capacitive humidity sensor for in-space manufacturing by electrohydrodynamic inkjet printing'. IEEE Sensors Journal. 2024;**24**(15):24659–24667. DOI: 10.1109/JSEN.2024.3400248.

[67] Ullah Z, *et al.* Facile assembly of flexible humidity sensors based on nanostructured graphite/zinc oxide-coated cellulose fibrous frameworks for human healthcare. RSC Advances. 2024;**14**(50):37570–37579. DOI: 10.1039/d4ra05761a.

[68] Strand EJ, *et al.* Ultrathin screen-printed plant wearable capacitive sensors for environmental monitoring. Advanced Sensor Research. 2025;**4**(3). DOI: 10.1002/adsr.202400177

[69] Somarathna US, *et al.* Reliability of screen-printed water-based carbon resistors for sustainable wearable sensors. IEEE Sensors Journal. 2025;**25**(4):6449–6463. DOI: 10.1109/JSEN.2024.3525055.

[70] Wadhwa A, *et al.* Flexible screen-printed sic-based humidity sensors. Communications Engineering. 2025;**4**(1). DOI: 10.1038/s44172-025-00425-2

[71] Aliane A, *et al.* Enhanced printed temperature sensors on flexible substrate. Microelectronics Journal. 2014;**45**(12):1621–1626. DOI: 10.1016/j.mejo.2014.08.011.

[72] Dankoco MD, Tesfay GY, Benevent E, Bendahan M. Temperature sensor

realized by inkjet printing process on flexible substrate. Materials Science and Engineering: B. 2016;**205**:1–5. DOI: 10.1016/j.mseb.2015.11.003.

[73] Vuorinen T, Niittynen J, Kankkunen T, Kraft TM, Mäntysalo M. Inkjet-printed graphene/PEDOT:PSS temperature sensors on a skin-conformable polyurethane substrate. Scientific Reports. 2016;**6**. DOI: 10.1038/srep35289.

[74] Wang C, Hong GY, Li KM, Young HT.A miniaturized nickel oxide thermistor via aerosol jet technology. Sensors (Switzerland). 2017;**17**(11). DOI: 10.3390/s17112602

[75] Shen A, Kim SB, Bailey C, Ma AWK, Dardona S. Direct write fabrication of platinum-based thick-film resistive temperature detectors. IEEE Sensors Journal. 2018;**18**(22):9105–9111. DOI: 10.1109/JSEN.2018.2869850

[76] Liu G, *et al.* A flexible temperature sensor based on reduced graphene oxide for robot skin used in internet of things. Sensors (Switzerland). 2018;**18**(5). DOI: 10.3390/s18051400

[77] Rivadeneyra A, *et al.* 'Cost-effective PEDOT : PSS temperature sensors inkjetted on a bendable substrate by a consumer printer'. Polymers (Basel). 2019;**11**(5). DOI: 10.3390/polym11050824

[78] Maskey BB, *et al.* A smart food label utilizing roll-to-roll gravure printed NFC antenna and thermistor to replace existing "use-by" date system. IEEE Sensors Journal [Internet]. 2020;**20**(4):2106–2116. DOI: 10.1109/JSEN.2019.2948752.

[79] Liew QJ, Aziz ASA, Lee HW, Lee MW, Hawari HF, Khir MHM. Inkjet-printed flexible temperature sensor based on silver nanoparticles Ink †. Engineering Proceedings. 2020;**2**(1). DOI: 10.3390/ecsa-7-08216

[80] Wang YF, *et al.* Fully printed PEDOT:PSS-based temperature sensor with high humidity stability for wireless healthcare monitoring. Scientific Reports. 2020;**10**(1). DOI: 10.1038/s41598-020-59432-2

[81] Chang YT, Hung KY, Young HT, Li KM, Chen RK. Aerosol jet printing of nickel oxide nanoparticle ink with ultraviolet radiation curing for thin-film temperature sensors. International Journal of Advanced Manufacturing Technology. 2022;**118**(5–6):1957–1965. DOI: 10.1007/s00170-021-08046-7

[82] Huber R, Eizaguirre S, Mescher H, Willig C, Zhang Q, Lemmer U.Fully printed temperature sensor array comprising 625 60×60 μm 2 pixels. Advanced Sensor Research. 2023;**2**(2). DOI: 10.1002/adsr.202200031

[83] Ketabi M, Shboul AA, Mahinnezhad S, Izquierdo R, 'Aerosol-jet printing of flexible green graphene humidity sensors for IoT applications', in *Proceedings of IEEE Sensors*, Institute of Electrical and Electronics Engineers Inc., 2021. DOI: 10.1109/SENSORS47087.2021.9639642.

[84] Shboul AA, *et al.* Graphene inks printed by aerosol jet for sensing applications: The role of dispersant on the inks' formulation and performance. Sensors. 2023;**23**(16). DOI: 10.3390/s23167151

[85] Tursunniyaz M, Meredith A, Andrews J. Aerosol jet printed resistive temperature sensors with high sensitivity. Sens Actuators A Physics. 2023;**364**. DOI: 10.1016/j.sna.2023.114777.

[86] Huang B, Wu S, Liu J, Liu J, Peng B, Zhou Z. Aerosol jet printing of polyelectrolyte-modified MXene ink for a multifunctional humidity and temperature flexible sensor. Chemical Engineering Journal. 2025;**519**. DOI: 10.1016/j.cej.2025.165403.

[87] Mallick MM, *et al.* High-sensitivity flexible thermocouple sensor arrays via printing and photonic curing. Advanced Functional Materials. 2024;**34**(20). DOI: 10.1002/adfm.202301681

[88] Hai Z, *et al.* Utilizing screen printing technology to fabricate tungsten-rhenium thick film thermocouples with a maximum temperature limit of 1600 °C. Measurement (Lond). 2025;**239**. DOI: 10.1016/j.measurement.2024.115454.

[89] Li S, *et al.* 'Screen printing silver shielded ITO–In2O3 Thin-Film thermocouples for electromagnetic interference reduction'. IEEE Sensors Letters. 2024;**8**(8):1–4. DOI: 10.1109/LSENS.2024.3423834.

[90] Soni S, Sathe P, Sarkar SK, Kushwaha A, Gupta D. Inkjet-printed sub-zero temperature sensor for real-time monitoring of cold environments. International Journal of Biological Macromolecules. 2024;**258**. DOI: 10.1016/j.ijbiomac.2023.128774.

[91] Wadhwa A, Benavides-Guerrero J, Gratuze M, Bolduc M, Cloutier SG. All screen printed and flexible silicon carbide NTC thermistors for temperature sensing applications. Materials. 2024;**17**(11):2489. DOI: 10.3390/MA17112489

[92] Miao Y, Dong H, Ma J, Wang Y, Jia Z, Tan Q. Fully printed direct-write preparation of high-temperature-resistant thermocouple sensors and their performance. Ceram International. 2025;**51**(3):3693–3702. DOI: 10.1016/j.ceramint.2024.11.345

[93] Turkani VS, *et al.* Nickel based RTD fabricated via additive screen printing process for flexible electronics. IEEE Access. 2019;**7**:37518–37527

[94] Mitra D, Mitra KY, Thalheim R, Zichner R.Inkjet-printed flexible thin-film thermal sensors for detecting elevated temperature range. Physica Status Solidi (A) Applications and Materials Science. 2024;**221**(4). DOI: 10.1002/pssa.202300562

[95] Tousignant MN, *et al.* Aerosol jet printed temperature sensors using an environmentally friendly bilayer dielectric. Flexible and Printed Electronics. 2024;**9**(1):015012. DOI: 10.1088/2058-8585/ad2ece.

[96] Corruccini RJ. Interpolation of platinum resistance thermometers, 20° to 273.15°K. Review of Scientific Instruments. 1960;**31**(6):637–640. DOI: 10.1063/1.1931274

[97] Cui J, *et al.* Fabrication and characterization of nickel thin film as resistance temperature detector. Vacuum. 2020;**176**:109288. DOI: 10.1016/J.VACUUM.2020.109288.

[98] Zhang Z, Tian B, Liu Y, Du Z, Lin Q, Jiang Z.Thermoelectric characteristics of silicon carbide and tungsten-rhenium-based thin-film thermocouples sensor with protective

coating layer by RF magnetron sputtering. Materials. 2019;**12**(12). DOI: 10.3390/MA12121981,

[99] Assumpcao D, Kumar S, Narasimhan V, Lee J, Choo H. High-performance flexible metal-on-silicon thermocouple. Scientific Reports. 2018;**8**(1):1–10. DOI: 10.1038/S41598-018-32169-9

[100] Wang H. Research on NiCr/NiSi thin film thermocouple sensor for measuring the surface temperature of automobile engine. Frontiers in Materials. 2025;**12**. 1554564. DOI: 10.3389/FMATS.2025.1554564.

[101] Nash JD, Caldwell DR, Zelman MJ, Moum JN. A thermocouple probe for high-speed temperature measurement in the ocean. Journal of Atmospheric and Oceanic Technology. 1999;**16**(11):1474–1482. DOI: 10.1175/1520-0426(1999)016<1474:ATPFHS>2.0.CO;2.

[102] He G, *et al.* La(Ca)CrO3-Filled SiCN Precursor Thin Film temperature sensor capable to measure up to 1100 °C High Temperature. Micromachines. 2023;**14**(9):1719. DOI: 10.3390/MI14091719.

[103] Fang Z, *et al.* Pt thin-film resistance thermo detectors with stable interfaces for potential integration in SiC high-temperature pressure sensors. Microsystems & Nanoengineering. 2024;**10**(1):1–12. DOI: 10.1038/S41378-024-00746-W.

[104] Qi Y, Wang J, Zhu Z, Li M. Thermo-reliability of PVD Cr/Au on sapphire substrates for high-temperature sensors. AIP Advances. 2022;**12**(6):65109. DOI: 10.1063/5.0088955/2819068.

[105] Jafarypouria M, Mahato B, Abaimov SG. Separating curing and temperature effects on the temperature coefficient of resistance for a Single-walled carbon nanotube nanocomposite. Polymers (Basel). 2023;**15**(2):433. DOI: 10.3390/POLYM15020433

[106] Eto H, *et al.* Design of sealable custom-shaped cell mimicries based on self-assembled monolayers on CYTOP Polymer. ACS Applied Materials & Interfaces. 2019;**11**(24):21372–21380. DOI: 10.1021/ACSAMI.9B05073/SUPPL_FILE/AM9B05073_SI_002.AVI.

[107] Selbmann F, *et al.* Impact of non-accelerated aging on the properties of Parylene C. Polymers (Basel). 2022;**14**(23):5246. DOI: 10.3390/POLYM14235246.

[108] Li W, *et al.* Microsecond-scale transient thermal sensing enabled by flexible Mo 1−x W x S 2 alloys . Research. 2024;**7**. DOI: 10.34133/RESEARCH.0452/SUPPL_FILE/RESEARCH.0452.F1.ZIP.

[109] Khuje S, *et al.* High-temperature oxidation-resistant printed copper conductors. Advanced Electronic Materials. 2023;**9**(3):2200979. DOI: 10.1002/AELM.202200979.

[110] Kim SW, *et al.* Encapsulation of polyvinyl alcohol based flexible temperature sensor through spatial atmospheric atomic layer deposition system to enhance its lifetime. Thin Solid Films. 2019;**673**:44–51

[111] Otomański P, Pawłowski E, Szlachta A. In Situ evaluation of the self-heating effect in resistance temperature sensors. Sensors (Basel). 2025;**25**(11):3374. DOI: 10.3390/S25113374

[112] Li J, Pei H, Kochan O, Wang C, Kochan R, Ivanyshyn A. Method for correcting error due to self-heating of resistance temperature detectors suitable for metrology in industry 4.0. Sensors (Basel). 2024;**24**(24):7991. DOI: 10.3390/S24247991

[113] Guan M, *et al.* Durable, breathable, sweat-resistant, and degradable flexible sensors for human motion detection. Chemical Engineering Journal. 2023;**462**:142151. DOI: 10.1016/J.CEJ.2023.142151.

[114] Zhang Y, *et al.* Highly stable flexible pressure sensors with a quasi-homogeneous composition and interlinked interfaces. Nature Communications. 2022;**13**(1):1–12. DOI: 10.1038/S41467-022-29093-Y.

[115] Xu R, *et al.* Lightweight flexible and efficient electromagnetic shielding composites based on silver-coated glass microspheres. ACS Omega. 2025;**10**(11):11516–11524. DOI: 10.1021/ACSOMEGA.5C00148/SUPPL_FILE/AO5C00148_SI_001.PDF.

[116] Blachowicz T, Wójcik D, Surma M, Magnuski M, Ehrmann G, Ehrmann A. Textile fabrics as electromagnetic shielding materials—A review of preparation and performance. Fibers. 2023;**11**(3):29. DOI: 10.3390/FIB11030029

[117] Janica I, *et al.* Covalently Functionalized MXenes for highly sensitive humidity sensors. Small Methods. 2023;**7**(8). DOI: 10.1002/SMTD.202201651,

[118] Li Z, *et al.* Asymmetrically patterned cellulose nanofibers/Graphene oxide composite film for humidity sensing and moist-induced electricity generation. ACS Applied Materials & Interfaces. 2020;**12**(49):55205–55214. DOI: 10.1021/ACSAMI.0C17970/SUPPL_FILE/AM0C17970_SI_004.AVI.

[119] Li Y, Huang X, Chen Q, Yao Y, Pan W. Nanochitin/MXene composite coated on quartz crystal microbalance for humidity sensing. Nanomaterials. 2023;**13**(24):3135. DOI: 10.3390/NANO13243135

[120] Xia J, Wang X, Wang X, Majer-Baranyi K, Zhang X. Hysteresis dynamic modeling and analysis of flexible Nano Silver-Polyvinyl alcohol humidity sensor based on the microscopic process and Langmuir-Fick Theory. ACS Omega. 2022;**7**(17):14994–15004. DOI: 10.1021/ACSOMEGA.2C00743,

[121] Füredi M, Alvarez-Fernandez A, Jesus M, Fornerod J, Fodor B, Guldin S. On the rational design of mesoporous Silica humidity sensors. Advanced Sensor Research. 2023;**2**(7):2200077. DOI: 10.1002/ADSR.202200077.

[122] Ma X, Zhang S, Zou P, Li R, Fan Y. Paper-based humidity sensor for respiratory monitoring. Materials. 2022;**15**(18):6447. DOI: 10.3390/MA15186447/S1

[123] Zhang X, Yang W, Zhang H, Xie M, Duan X. PEDOT:PSS: From conductive polymers to sensors. Nanotechnology and Precision Engineering. 2021;**4**(4):45004. DOI: 10.1063/10.0006866/626875.

[124] Wang C, Jiao C, Wang M, Pan J, Wang Q. GO/CNT-OH/Nafion nanocomposite humidity sensor based on the LC wireless method. Nanomaterials. 2023;**13**(13):1925. DOI: 10.3390/NANO13131925/S1

[125] Zhu P, *et al.* Flexible and Highly Sensitive Humidity Sensor Based on Cellulose Nanofibers and Carbon Nanotube Composite Film. Langmuir. 2019;**35**(14):4834–4842. DOI: 10.1021/ACS.LANGMUIR.8B04259.

[126] 'Screen printing technology for energy devices - andreas willfahrt - Google Libros'. [Online]. Available: https://books.google.ca/books?id=BM11DwAAQBAJ. [Accessed: 2025-May-27]

[127] Liang T-X, Sun WZ, Wang L-D, Wang YH, Li H-D. Effect of surface energies on screen printing resolution. IEEE Transactions on Components, Packaging, and Manufacturing Technology: Part B. 1996;**19**(2):423–426. DOI: 10.1109/96.496047

[128] Liu T, Tiu C, Chen L, and Liu D. The Influence of Slurry Rheology on Lithium-ion Electrode Processing. In Printed Batteries. Lanceros-Méndez S, and Costa CM. editors. 1st. Chichester, UK: Wiley; 2018. pp. 63–79. DOI: 10.1002/9781119287902.ch3

[129] Potts S-J, Phillips C, Claypole T, Jewell E. The effect of carbon ink rheology on ink separation mechanisms in screen-printing. Coatings. 2020;**10**(10):1008. DOI: 10.3390/coatings10101008

[130] Liu Y, Zhu H, Xing L, Bu Q, Ren D, Sun B. Recent advances in inkjet-printing technologies for flexible/wearable electronics. Nanoscale. 2023;**15**(13):6025–6051. DOI: 10.1039/D2NR05649F

[131] Sajedi-Moghaddam A, Rahmanian E, Naseri N. Inkjet-printing technology for supercapacitor application: Current state and perspectives. ACS Applied Materials & Interfaces. 2020;**12**(31):34487–34504. DOI: 10.1021/ACSAMI.0C07689,

[132] Li C, Bu F, Wang Q, Liu X. Recent developments of inkjet-printed flexible energy storage devices. Advanced Materials Interfaces. 2022;**9**(34):2201051. DOI: 10.1002/ADMI.202201051.

[133] Liu Y-F, Tsai M-H, Pai Y-F, Hwang W-S. Control of droplet formation by operating waveform for inks with various viscosities in piezoelectric inkjet printing. Applied Physics A. 2013;**111**(2):509–516. DOI: 10.1007/s00339-013-7569-7

[134] Wang S, Zhong Y, Fang H. Deformation characteristics of a single droplet driven by a piezoelectric nozzle of the drop-on-demand inkjet system. Journal of Fluid Mechanics. 2019;**869**:634–645. DOI: 10.1017/jfm.2019.242.

[135] Elsersawy R, Rahman A, Sakib-Uz-Zaman C, Abu M, Khondoker H. Multifunctional inks in aerosol jet printing: Performance, challenges, and applications. Frontiers in Manufacturing Technology. 2025;**5**. 1558209. DOI: 10.3389/FMTEC.2025.1558209.

[136] Deiner LJ, Reitz TL. Inkjet and aerosol jet printing of electrochemical devices for energy conversion and storage. Advanced Engineering Materials. 2017;**19**(7):1600878. DOI: 10.1002/ADEM.201600878.

[137] Bappy MO, Jiang Q, Atampugre S, Zhang Y, 'Aerosol jet printing of high-temperature multimodal sensors for strain and temperature sensing', 2024 [Online]. Available: https://arxiv.org/pdf/2403.01524 [Accessed: 2025-July-28]

[138] Yang C, *et al.* Roll-to-roll prelithiation of lithium-ion battery anodes by transfer printing. Nature Energy. 2023;**8**(7):703–713. DOI: 10.1038/s41560-023-01272-1.

[139] Patil GC. Doctor blade: A promising technique for thin film coating. Simple Chemical Methods for Thin Film Deposition: Synthesis and Applications. 2023;509–530. DOI:10.1007/978-981-99-0961-2_12

[140] Assaifan AK. Flexographic Printing Contributions in Transistors Fabrication. Advanced Engineering Materials. 2021;**23**(5):2001410. DOI: 10.1002/ADEM.202001410.

[141] Morais M, *et al.* Flexographic printed microwave-assisted grown zinc oxide nanostructures for sensing applications. Materials Horizons. 2024;**11**(24):6463–6475. DOI: 10.1039/D4MH01000K.

[142] Morais M, *et al.* Flexographic printed flexible thermochromic stickers for smart sensing applications. Advanced Materials Interfaces. 2025;**12**(11):2400972. DOI: 10.1002/ADMI.202400972.

[143] Manapat JZ, Chen Q, Ye P, Advincula RC. 3D printing of polymer nanocomposites via stereolithography. Macromolecular Materials and Engineering. 2017;**302**(9):1600553. DOI: 10.1002/MAME.201600553;WGROUP:STRING:PUBLICATION.

[144] Ertugrul I, Ulkir O, Ersoy S, Ragulskis M. Additive manufactured strain sensor using stereolithography method with photopolymer material. Polymers. 2023;**15**(4):991. DOI: 10.3390/POLYM15040991

[145] Massaroni C, Saroli V, Aloqalaa Z, Presti DL, Schena E. Extrusion-based fused deposition modeling for printing sensors and electrodes: Materials, process parameters, and applications. Smart Mathimatics. 2025;**6**(4):e70027. DOI: 10.1002/SMM2.70027

[146] Sathies T, Senthil P, Anoop MS. A review on advancements in applications of fused deposition modelling process. Rapid Prototyping Journal. 2020;**26**(4):669–687. DOI: 10.1108/RPJ-08-2018-0199

[147] Stefano JS, *et al.* Electrochemical (Bio)Sensors enabled by fused deposition modeling-based 3D printing: A guide to selecting designs, printing parameters, and post-treatment protocols. Analytical Chemistry. 2022;**94**(17):6417–6429. DOI: 10.1021/ACS.ANALCHEM.1C05523.

[148] Li D, Lai WY, Feng F, Huang W. Post-treatment of screen-printed silver nanowire networks for highly conductive flexible transparent films. Adv Mater Interfaces. 2021;**8**(13):2100548. DOI: 10.1002/ADMI.202100548.

[149] Kantaros A, Ganetsos T, Petrescu FIT, Ungureanu LM, Munteanu IS. Post-production finishing processes utilized in 3D printing technologies. Processes. 2024;**12**(3):595. DOI: 10.3390/PR12030595

[150] Abdullah AN, *et al.* Correction model for metal oxide sensor drift caused by ambient temperature and humidity. Sensors (Basel). 2022;**22**(9):3301. DOI: 10.3390/S22093301.

[151] Ali I, *et al.* A highly accurate, polynomial-based digital temperature compensation for piezoresistive

pressure sensor in 180 nm CMOS technology. Sensors. 2020;**20**(18):5256. DOI: 10.3390/S20185256.

[152] Pérez-Bailón J, Calvo B, Medrano N. A CMOS low pass filter for SoC Lock-in based measurement devices. Sensors (Basel). 2019;**19**(23):5173. DOI: 10.3390/S19235173

[153] Panetas-felouris O, Vlassis S. A 3rd-order FIR Filter Implementation Based on time-mode signal processing. Electronics. 2022;**11**(6):902. DOI: 10.3390/ELECTRONICS11060902

[154] Tanaka T, Nambu I, Maruyama Y, Wada Y. Sliding-Window normalization to improve the performance of machine-learning models for Real-Time motion prediction using electromyography. Sensors (Basel). 2022;**22**(13):5005. DOI: 10.3390/S22135005

[155] Yin Dong Z, Lin Xu J. Baseline estimation using optimized asymmetric least squares (O-ALS). Measurement. 2024;**233**. 114731. DOI: 10.1016/J.MEASUREMENT.2024.114731.

[156] Zhang S, Dai Q, 'Nonlinear calibration of thermocouple sensor based on least squares support vector regression', *Proceedings of the 2015 5th International Conference on Computer Sciences and Automation Engineering*, vol. **42**, pp. 5–10. 2016, DOI: 10.2991/ICCSAE-15.2016.2.

[157] Khodarahmi M, Maihami V. A review on kalman filter models. Archives of Computational Methods in Engineering. 2023;**30**(1):727–747. DOI: 10.1007/s11831-022-09815-7

[158] Gabr K, Abdelkader M, Jarraya I, AlMusalami A, Koubaa A, 'SMART-TRACK: A Novel Kalman Filter-Guided Sensor Fusion For Robust UAV Object Tracking in Dynamic Environments', 2024 [Online]. Available: https://arxiv.org/pdf/2410.10409 [Accessed: 2025-July-29]

[159] Andrade P, Silva I, Silva M, Flores T, Cassiano J, Costa DG. A Tiny ML soft-sensor approach for low-cost detection and monitoring of vehicular emissions. Sensors. 2022;**22**(10):3838. DOI: 10.3390/S22103838

[160] Zhuo S, *et al.* An empirical study of low precision quantization for TinyML. Mar. 2022. DOI: 10.48550/arXiv.2203.05492.

[161] Aymone FM, Pau DP. Benchmarking in-sensor machine learning computing: An Extension to the MLCommons-Tiny suite. Information 2024. 2024;**15**(11):674. DOI: 10.3390/INFO15110674

[162] Darwish T, Bayoumi M. Trends in Low-Power VLSI Design. The Electrical Engineering Handbook. 2005;263–280. DOI:10.1016/B978-012170960-0/50022-0

[163] Du W, Jia X, Cheng Z, Xu W, Li Y, Dai L. Low-power-consumption CMOS inverter array based on CVD-grown p-MoTe2 and n-MoS2. iScience. 2021;**24**(12):103491. DOI: 10.1016/J.ISCI.2021.103491.

[164] Liu FT, Ting KM, Zhou ZH, 'Isolation forest', *Proceedings - IEEE International Conference on Data Mining, ICDM*, pp. 413–422, 2008, DOI: 10.1109/ICDM.2008.17.

Chapter 2

Fiber Optics Technology for Temperature and Humidity Sensor Development

Renato Faraco Filho, João Gabriel de Carvalho, Luiz David Silva, Enzo Dantas da Silva, Sinei Gonçalves Júnior, Flávio Augusto Marques, Alexandre Cotta, Alexandre Bessa dos Santos, Jonas Henrique Osório and Jefferson E. Tsuchida

Abstract

This chapter reviews optical fiber sensors designed for measuring temperature and humidity, covering their fundamental principles, advanced architectures, and functionalized designs. The review discusses several sensor platforms, including those based on fiber Bragg gratings (FBGs), Long-Period gratings (LPGs), and Fabry–Perot interferometers (FPIs), as well as multimode interference (MMI) sensors, microstructured fibers, and distributed fiber sensors, describing the sensing mechanisms and the corresponding working principles. Additionally, the chapter presents strategies to enhance sensitivity and selectivity through microstructural modifications, material infiltration, and coating techniques with materials such as graphene oxide (GO) and polymers. This work demonstrates that fiber-based sensors offer high versatility for temperature and humidity monitoring, which is highly beneficial for industrial, environmental, and biomedical applications. This family of devices, hence, opens new avenues for sensitive, compact, and fast-response platforms capable of supporting the development of novel sensing technologies.

Keywords: optical fiber sensors, fiber Bragg gratings, long-period gratings, interferometric sensors, multimodal interference sensors, microstructured fibers, functional coatings, distributed fiber sensors

1. Introduction

Optical fiber sensors have emerged as versatile and highly sensitive platforms for monitoring a wide range of physical, chemical, and environmental parameters. Their advantages, including immunity to electromagnetic interference, compactness, and the ability to operate in harsh environments, have allowed the identification of this family of sensors as eligible candidates for industrial, biomedical, and environmental applications.

To fully capitalize on these benefits, a diverse array of advanced fiber sensor designs and architectures has been developed. These innovations move beyond simple light transmission, using specialized optical and material properties to actively modulate signals in response to external stimuli. The engineering of these sophisticated systems enables the high-performance and multiparameter sensing capabilities of fiber devices, making them highly valuable sensing platforms.

Thus, this chapter provides an overview of advanced optical fiber sensing technologies for temperature and humidity sensing, focusing on fiber Bragg gratings (FBGs), long-period fiber gratings (LPFGs), microstructured fibers, antiresonant fibers, and functionalized fibers. The discussion emphasizes the devices' working principles and relevant demonstrations in the literature, as well as a description of the sensors' performance parameters.

Furthermore, the chapter examines strategies to enhance the devices' sensitivity and selectivity, including the use of microstructural design, liquid infiltration, and functional coatings. By analyzing recent literature and experimental results, the chapter highlights both the strengths of each approach, offering guidance for selecting the most suitable technology based on application requirements. Ultimately, this work aims to provide a framework for understanding current optical fiber temperature and humidity sensors and their potential for future developments in sensing technology.

2. Fiber Bragg grating sensors

Fiber Bragg gratings (FBGs) represent one of the most mature technologies in fiber optic sensing, enabling the development of systems capable of monitoring different parameters, including temperature [1], strain [2], humidity [3], vibration [4], and pressure [5]. Indeed, since FBGs' early demonstrations, FBG sensors have garnered great interest due to their low insertion loss, narrowband wavelength reflection, linearity, and compatibility with fiber optics infrastructure [6]. An FBG is a periodic modulation of the refractive index in the optical fiber core, which results in a characteristic reflection spectrum. This spectrum is defined by a narrow peak centered at the Bragg wavelength (λ_B), which is given by Eq. (1), where n_{eff} is the effective refractive index of the fiber core's fundamental mode and Λ is the grating pitch. Particularly, variations in external parameters such as temperature or strain can entail a shift in the Bragg wavelength, forming the basis for the realization of sensing measurements [7]. **Figure 1** illustrates an FBG, indicating the reflection of a spectral peak centered at λ_B

$$\lambda_B = 2n_{eff}\Lambda \tag{1}$$

The Bragg grating wavelength shift caused by variations in temperature and strain is given by Eq. (2) [8], where α is the thermal expansion coefficient of the fiber, ξ is the thermo-optic coefficient, ΔT is the temperature variation, p_e is the effective photoelastic constant, and ε is the applied strain. This expression indicates that both thermal and mechanical effects contribute to the spectral response of fiber Bragg gratings, which makes them highly versatile but also requires compensation schemes in multi-parameter sensing applications [9]. In particular, we note that the temperature effect on a Bragg grating is represented by the first term in Eq. (2). The Bragg wavelength hence shifts as

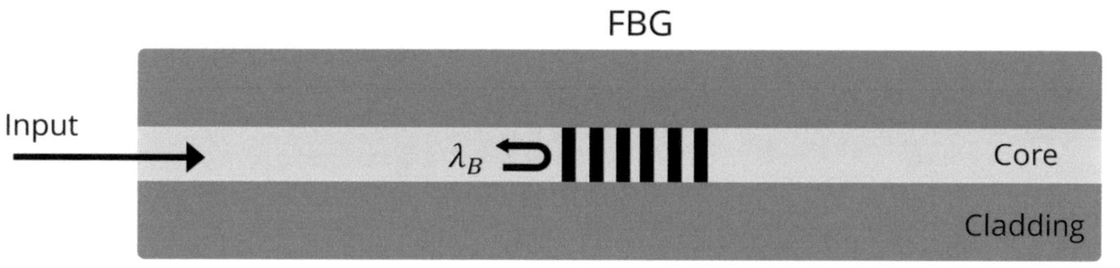

Figure 1.
Illustration of an FBG, indicating that the forward-propagating core mode is coupled to a back-propagating core mode at λ_B.

a result of thermal expansion-induced modifications to the grating pitch and the thermo-optic effect.

$$\Delta\lambda_B = \lambda_B[(\alpha + \xi)\Delta T + (1 - p_e)\varepsilon], \tag{2}$$

Indeed, temperature sensing is one of the most typical applications of FBGs. Consequently, several configurations have been demonstrated to date, both with regard to the FBG acting as the temperature sensor itself or as a tool for multiparameter sensing [10] and biomedical materials characterization [11]. Furthermore, it should be mentioned that researchers have recently dedicated their efforts to the development of new configurations that minimize cross-sensitivity with strain effects [12]. Additionally, they have explored the use of metallic coatings to enhance sensitivity [13] and have conducted studies on the temporal response of FBGs to temperature variations [14]. Moreover, efforts have been made to develop multipoint temperature sensors [15].

Additionally, besides being effective for measuring strain and temperature, FBGs also play an important role in humidity sensing. This application can be enabled by functionalizing the fiber with hygroscopic coatings sensitive to moisture. When the surrounding humidity changes, the coating material swells or shrinks, which, in turn, applies mechanical strain to the optical fiber, causing a measurable shift in the Bragg wavelength [3]. Additionally, it is possible to use tilted Bragg gratings (TFBGs), which are inscribed at an angle to the fiber's core. The tilted structure couples light to both core and cladding modes, allowing for an effective assessment of the sensitivity of the cladding modes to the surrounding environment and enabling temperature cross-sensitivity mitigation [16]. Regarding the humidity-sensitive coatings, several materials can be used, including biopolymers such as chitosan [17] and synthetic polymers. For instance, studies have shown that a polyimide-coated FBG exhibits a humidity sensitivity ranging from 1.3 to 1.5 pm/%RH with good linearity [3]. Other approaches utilize materials such as graphene oxide (GO), which shows high hydrophilicity and represents a promising material for the development of humidity fiber sensors [18].

3. Long-Period Grating sensors

Long-period fiber gratings (LPG) are devices that couple the core mode with co-propagating cladding modes. This is typically achieved through refractive index or geometric periodic modulations along the optical fiber axis on the order of hundreds of micrometers [19]. The strong attenuation experienced by the forward-propagating cladding mode causes a change in the transmitted spectrum, which

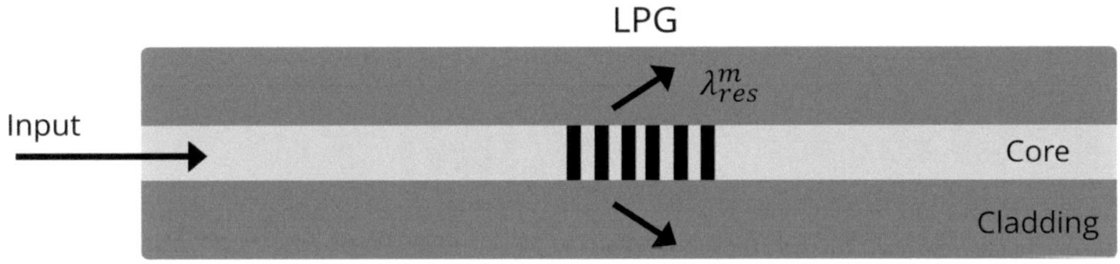

Figure 2.
Illustration of an LPG, indicating that the grating couples the core mode to cladding modes at the resonant wavelengths λ^{m}_{res}.

exhibits a series of attenuation bands, each centered at a specific wavelength, λ^{m}_{res}corresponding to the coupling of different cladding modes [20]. The spectral positions of the attenuation bands can be calculated by Eq. (3), where $n_{eff,co}$ and $n^{m}_{eff,cl}$ are the effective refractive indices of the core and cladding modes, and Λ is the grating pitch. **Figure 2** represents an LPG, indicating that the grating couples the core mode to cladding modes at the resonant wavelengths λ^{m}_{res}.

$$\lambda^{m}_{res} = (n_{eff,co} - n^{m}_{eff,cl}).\Lambda \tag{3}$$

The sensitivity of long-period grating spectral features is influenced by the period, the characteristics of the surrounding medium (which impacts $n^{m}_{eff,cl}$sensitivity and the order of the cladding coupling mode. Furthermore, the materials present in the optical fiber composition also influence the sensitivity parameters [21]. Thus, many physical, chemical, and biological parameters can induce changes in LPGs' characteristic spectrum, motivating the development of a wide array of optical fiber sensors.

In addition to thermal sensitivity, LPFGs can be used as relative humidity (RH) sensors, taking advantage of the modification of the refractive index of the surrounding medium. A typical strategy involves the deposition of hygroscopic polymer coatings on the grating region, so that the absorption of water molecules by the coating alters its effective refractive index, causing shifts in the resonant wavelengths [22].

In this context, researchers have employed materials such as polyvinyl alcohol (PVA), whose high affinity for water makes it suitable for this application. When the PVA coating comes into contact with higher humidity environments, a swelling process occurs, simultaneously modifying its physical and optical properties. As a consequence, a spectral shift of the LPFG attenuation bands is observed, proportional to the RH variation [23]. Experiments conducted between 33% RH and 97% RH showed that LPFG-based sensors coated with PVA exhibited response times on the order of tens of seconds and measurable shifts in the attenuation dips, demonstrating the feasibility of the technique.

In addition to PVA, other coatings have been explored to increase sensor selectivity and stability. Zheng et al. [24] demonstrated the use of multilayer nanofilms obtained by electrostatic self-assembly, in which a first layer increases the overall sensitivity to the refractive index, while a second oxide-based coating with a compatible structure is responsible for selectivity to water molecules. This approach resulted in sensors capable of detecting subtle humidity variations in porous materials with good thermal stability and potential applications in structural monitoring.

Additionally, graphene oxide (GO) has also been used in the context of LPG sensors. GO exhibits a unique two-dimensional structure, resulting in faster response times and higher sensitivity compared to alternative materials. Tsai et al. [25] proposed an S-Type LPG coated with GO for relative humidity sensing, using microelectromechanical processes and inkjet piezoelectric printing technology. In this demonstration, a single-mode optical fiber was etched to different diameters, and the sensing experiments demonstrated that the S-Type LPG coated with GO exhibited high sensitivity to relative humidity variations. Additionally, the sensor demonstrated good linearity, high stability, and reversibility over a wide operating range.

4. Fabry–Perot interferometer sensors

The Fabry–Perot interferometer (FPI) is an important optical device with extensive applications, particularly when integrated with optical fiber technology. An FPI operates on the principle of optical interference, using a cavity formed between two optical reflectors. Light entering the cavity is reflected back and forth, and the superposition of these reflected signals generates an interference pattern. The characteristics of this pattern are highly sensitive to changes in the optical path length of the cavity [26].

Fiber Fabry–Perot interferometers (FPIs) are generally classified into two categories: intrinsic and extrinsic. An intrinsic FPI (IFPI) typically employs in-line reflectors formed within the fiber itself, a process that often requires specialized micromachining. In contrast, an extrinsic FPI (EFPI) creates its cavity externally, typically as an air gap between the end of the optical fiber and a separate external reflector. This design offers greater flexibility in sensor construction, as it typically does not require high-precision micromachining to create the cavity. **Figure 3** illustrates IFPI and EFPI structures.

The reflection from a low-finesse EFPI can be modeled as a two-beam interference pattern, as expressed in Eq. (4), where I is the total reflected light intensity, I_1 and I_2 are the reflected light intensities from the two reflectors, respectively, φ_0 is a phase shift induced by the reflectors, and φ is the round-trip phase delay given by Eq. (5), where n is the refractive index of the cavity medium, L is the physical length of the cavity, and λ is the light wavelength. A change in the cavity length or in the refractive index of the cavity medium due to an external perturbation directly alters the interference signal, allowing the device to act as a sensor [26].

$$I = I_1 + I_2 + 2\sqrt{I_1 I_2 cos(\varphi + \varphi_0)} \tag{4}$$

$$\varphi = \frac{4\pi nL}{\lambda} \tag{5}$$

While EFPIs are widely used for mechanical and physical sensing by correlating an external perturbation to a change in the cavity's physical length, they can also be specifically adapted for temperature and humidity sensing. Regarding their temperature dependence, FPIs based on air cavities can be set to be inherently

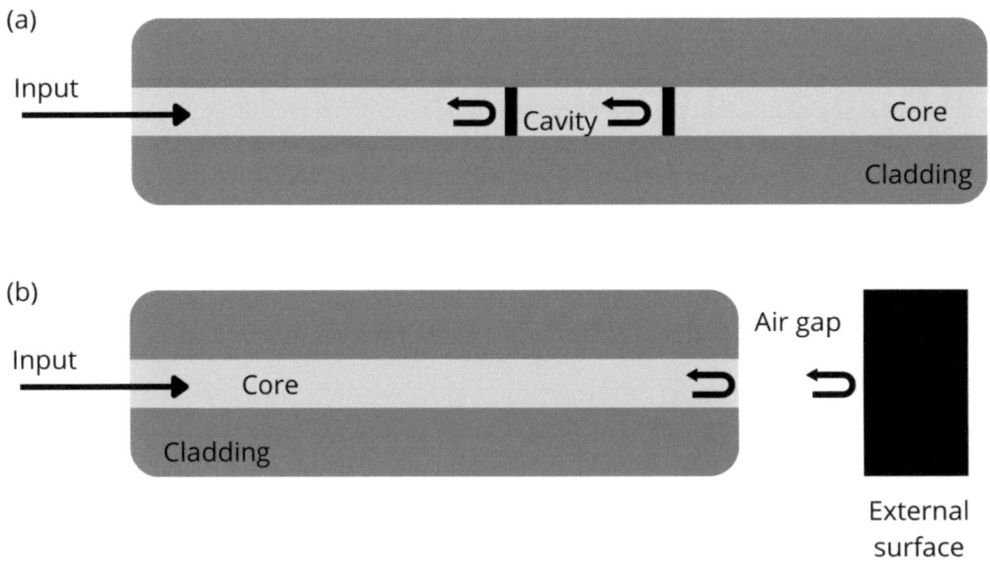

Figure 3.
Illustration of (a) an IFPI and (b) an EFPI. In an IFPI, the reflectors are created within the fiber, and in an EFPI, the cavity is created outside the optical fiber.

insensitive to temperature fluctuations, offering an advantage over other fiber-optic sensors such as FBGs that display significant temperature cross-sensitivity. To create a dedicated temperature sensor, a material with large thermal expansion or thermo-optic coefficients can be integrated into the cavity.

While the air-gap cavity in EFPI can provide inherent immunity to temperature-induced crosstalk, this characteristic makes the platform promising for high-temperature sensing when integrated with an appropriate transducer. To achieve this, researchers have focused on approaches such as employing materials with excellent thermal stability to obtain the EFPI's diaphragm, including ceramic materials [27] and sapphire [28], pushing the operating temperature limits beyond 700°*C*. These high-performance sensors are often fabricated using advanced methods like femtosecond laser micromachining or multistep arc discharge techniques. Recent works have also combined the EFPI principle with microelectromechanical systems fabrication, creating batch-producible sensors that can withstand temperatures up to 800°*C* [29].

EFPIs can also be employed for humidity sensing, often using a similar structure to the diaphragm-based pressure sensor. In this application, a thin layer of moisture-sensitive material, such as chitosan [30] and Nafion [31], is deposited on the tip of an optical fiber. In this application, the sensor's operation relies on the swelling effect of the chitosan film as it absorbs moisture. This swelling changes the length of the Fabry–Perot cavity, which is then accounted for to determine the humidity level. Other materials, including graphene oxide [32] and polyimide [33], have also been explored to improve sensor performance, achieving fast responses and high sensitivity.

5. Multimode interference sensors

Multimode interference (MMI) sensors also motivate important research regarding the development of temperature and humidity sensors. A typical MMI sensing platform is the singlemode-multimode-singlemode (SMS) fiber structure, which is built by splicing a short section of a multimode fiber (MMF) between two singlemode fibers (SMFs). This simple structure operates based on the MMI principle and typically explores the self-imaging phenomenon. As light from the input SMF enters the MMF section, it excites multiple modes that propagate independently. These modes interfere with each other, leading to a reproduction of the input light field profile, known as self-imaging. The output spectrum is highly sensitive to changes in the surrounding environment, which alter the modal interference pattern [34].

The sensitivity of SMS structures to external perturbations makes them suitable for a wide range of sensing applications, including temperature and humidity monitoring. In applications based on the self-imaging phenomenon, a spectral transmission peak is assessed during the measurements. Specifically, this peak occurs at a wavelength λ_0 when the length of the MMF (L_{MMF}) corresponds to a self-imaging distance (Z_{SI}), that is, the situation in which the light field from the input fiber is replicated. This relationship can be expressed by Eq. (6), where n_{core} is the refractive index of the fiber core and D is the diameter of the core mode in a high-contrast waveguide [35]. **Figure 4** represents the SMS configuration, indicating the coupling of the SMF core mode to multiple core modes in the MMF, resulting in filtering of the input spectrum around λ_0 due to the self-imaging phenomenon.

$$Z_{\mathrm{SI}} = \frac{4n_{\mathrm{core}}D_{\mathrm{core}}^2}{\lambda_0} \tag{6}$$

For temperature sensing, shifting of the characteristic transmission peak can be monitored as the effective refractive index of the optical modes involved and the dimensions of the fiber change with temperature. Examples include traditional SMS sensors, SMS sensors with graded-index multimode fibers, bent SMS sensors, and configurations using hollow-core fibers [34]. A successful approach for obtaining temperature SMS sensors is to use singlemode-no-core-singlemode and explore their

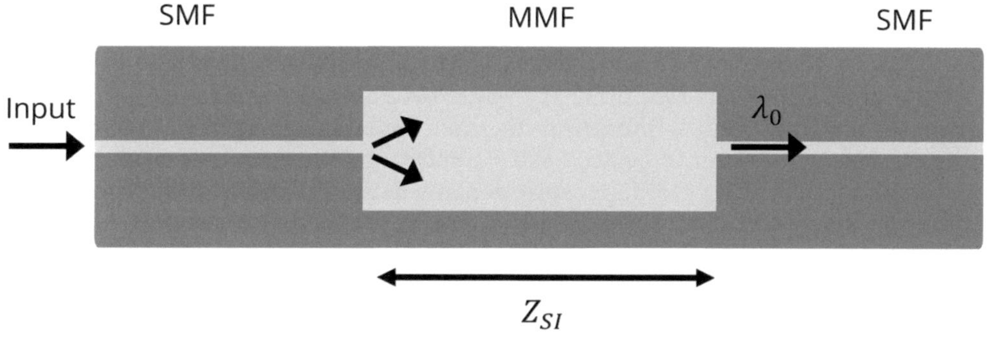

Figure 4.
Illustration of an SMS configuration, indicating the coupling of the SMF core mode to multiple core modes in the MMF, resulting in filtering of the input spectrum around λ_0 due to the self-imaging phenomenon.

refractive index sensitivity, with the sensor operation being mediated by the thermo-optic effect [36].

For humidity sensing, these sensors typically rely on a moisture-sensitive material that coats the fiber. The change in humidity causes the coating to swell or change its refractive index, which, in turn, impacts the multimode interference pattern. Examples include sensors based on thin-core fibers coated with methylcellulose films [37] and sensors utilizing advanced materials such as carbon-nanotube/polyvinyl alcohol (CNT/PVA) [38]. Additionally, a high-temperature humidity sensor based on a side-polished multimode fiber coated with gelatin was also developed [39].

6. Microstructured optical fiber sensors

Microstructured optical fibers (MOFs) are a type of optical fiber that encompass micrometric features (typically air holes) on their cross-section that run along their length, giving them unique optical properties typically inaccessible in conventional fibers. These structural characteristics allow for unprecedented control over the fiber's light-guiding properties, making them highly adaptable for a myriad of applications. Because of their design flexibility, MOFs have become a versatile platform for developing a vast variety of sensors. These sensors can measure a wide range of parameters, including temperature, humidity, refractive index, pressure, and curvature [40].

MOFs provide an effective platform for temperature sensing due to the wide design possibilities and the ability to integrate temperature-sensitive materials. Unlike conventional fibers, whose temperature sensitivity is limited by the thermo-optic coefficient difference between the core and cladding, MOF-based sensors can achieve high sensitivity by infiltrating their air holes with materials that have large thermo-optic coefficients [41]. For instance, a high-sensitivity temperature sensor was developed by filling the air holes of a high-birefringence photonic crystal fiber (HiBi-PCF) with alcohol [42]. This sensor, configured as a fiber loop mirror, achieved a temperature sensitivity of 6.6 nm/°C by measuring the shift of resonant dips caused by the refractive index change of the alcohol.

The thermal expansion of metal fillers can also be used for obtaining MOF-based temperature sensors. For instance, a Sagnac loop interferometer with a side-hole PCF filled with indium achieved a very high sensitivity of −9.0 nm/°C [43]. Additionally, metal-filled capillary embedded-core fibers emerged as promising platforms for temperature sensing, showing sensitivities as high as 8.4 nm/°C [44]. In these applications, expansion of the metal inside the fiber induces mechanical stress within its structure, impacting the fiber modal birefringence. This birefringence variation can be accessed via spectral polarimetric measurements, enabling the realization of temperature sensing demonstrations. **Figure 5a** and **Figure 5b** illustrate the cross-section of a side-hole PCF and of a capillary embedded-core fiber.

Additionally, in recent years, hollow-core optical fibers (HCF) have attracted growing interest due to the development of new fiber designs and impressive loss-decreasing demonstrations [45]. In this type of special fiber, the core has a refractive index lower than that of the cladding, and light is confined thanks to the inhibited coupling guidance mechanism (the literature also refers to this as the antiresonant

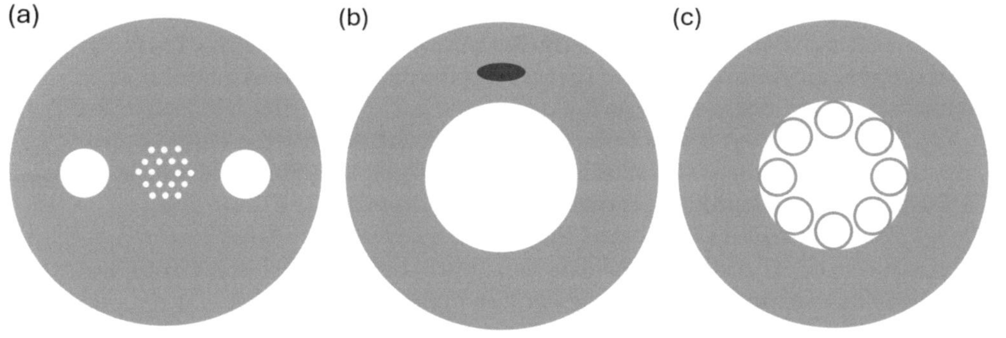

Figure 5.
Illustration of (a) a side-hole photonic crystal fiber, (b) a capillary embedded-core fiber, and (c) a tubular hollow-core fiber. Illustration of (a) a side-hole photonic crystal fiber, (b) a capillary embedded-core fiber, and (c) a tubular hollow-core fiber.

guiding mechanism). The transmission spectrum of this fiber type alternates between regions of high and low loss depending on the waveguide construction parameters and the wavelength.

The parameters that determine the spectral positions of the high-attenuation regions are established by Eq. (7), where t is the cladding tube's thickness, n_1 and n_2 are the refractive indices of the core and the cladding material, respectively, and m is an integer identifying the resonance order.

$$\lambda_m = \frac{2t}{m}\sqrt{n_2^2 - n_1^2} \tag{7}$$

The hollow-core fiber family includes capillary fibers and hollow-core microstructured fibers, both generally made from silica. For optical transmission to occur via the antiresonant guiding mechanism, it is essential that the core refractive index is lower than that of the fiber walls. In this context, several optical sensors have been developed by inserting materials into hollow-core optical fibers. Among these, one study proposed a high-sensitivity, fast-response temperature sensor based on the antiresonant guiding mechanism [46]. The device was configured according to an SMF–HCF–SMF structure and, to increase the interaction of the guided light with the sensing medium, the hollow-core fiber was side-polished, partially exposing the core, which was then filled with polydimethylsiloxane (PDMS). This polymer, with a refractive index lower than silica, enabled guidance via the antiresonance condition. Moreover, its high thermo-optic coefficient caused the refractive index to vary significantly with temperature, leading to shifts in the transmission bands. The results obtained showed that the sensor achieved a sensitivity of 4.223 nm/°C for a compact length 1.6mm, along with a fast response time of 16 ms, in the temperature range of 25–85°C. Stability tests conducted over three months evidenced high reversibility and accuracy, confirming the reliability of the sensor [46].

Similarly to the previously mentioned work, low-refractive-index liquids were introduced into the interior of a silica capillary to develop an optical temperature sensor using the SMF–silica capillary–SMF structure. Internal filling and external immersion in liquids with refractive indices lower than silica enabled light confinement by the antiresonant guiding mechanism, as internal reflection occurs at the silica–liquid interfaces. The presence of liquids in both regions changes the resonance conditions, promoting shifts in the resonant wavelengths due to variations in refractive index caused by the thermo-optic coefficient of the liquid. Thus, an increase in temperature changes the refractive index, resulting in spectral band shifts. Experimentally, it was found that thermal sensitivity depended on the refractive index of the infiltrated liquid, reaching up to 2.99 nm/°C in the range of 25–60 °C for liquids with an RI of 1.39 [47].

Additionally, studies have investigated the insertion of materials into microstructured HCF, with emphasis on tubular fibers. These structures have the advantage of functioning simultaneously as microreservoirs and waveguides, enabling direct interaction between the inserted material and the propagated optical mode. Among the reported strategies, we mention the use of water as the filling medium, taking advantage of its high thermo-optic coefficient compared to that of silica. In this study, such a configuration resulted in a sensitivity of 0.42 nm/°C, in the range of 41 to 71 °C, demonstrating the potential of the platform for temperature sensing applications [48]. In a subsequent study, partial insertion of materials into the hollow-core microstructured fiber was proposed, allowing for the design of a temperature sensor with improved resolution levels [49].

MOFs have also been used for humidity sensing by designing their structures to interact with the environment, often by integrating hygroscopic materials. These materials change their refractive index and/or volume in response to humidity, which, in turn, affects the light propagating through the fiber, enabling sensing measurements. For example, authors have demonstrated a humidity sensing platform based on a suspended tri-core MOF coated with a chitosan polymer matrix [50]. This fiber has three small cores suspended by thin silica struts inside larger air holes, allowing for strong interaction between the light's evanescent field and the chitosan in the air holes. The chitosan acts as a humidity-sensitive material because its RI and volume change as it absorbs water vapor, which causes a shift in the wavelength of the reflected light. An FPI is created by depositing the chitosan film in the MOF's air holes, serving as the sensing platform. This sensor demonstrated a maximum sensitivity of 81.05 pm/%RH in the 90–95% relative humidity (RH) range and exhibited a very fast response with a rise time of 80 ms and a recovery time of 70 ms. The sensor's fast response makes it suitable for real-time human breath monitoring [50].

Another example of using MOFs as humidity sensors employs a photonic crystal fiber (PCF) coated with graphene oxide [51]. The sensor is based on a Mach–Zehnder interferometer (MZI), where two spherical structures act as couplers to excite and re-couple different modes between the SMF and the PCF. Graphene oxide is a promising coating material because of its hydrophilic properties, meaning it readily absorbs water molecules from the surrounding environment. This absorption causes a change in the refractive index of the GO film, which, in turn, modulates the effective RI of the cladding modes in the PCF. This change is detected as a shift in the interference spectrum, allowing for humidity measurement. This sensor was tested in a range from 30% to 70% RH and achieved sensitivities of 0.128 nm/%RH and 0.159 nm/%RH for its two interference dips, demonstrating a significant improvement in sensitivity after coating with GO compared to the uncoated fiber. The simple fabrication and high sensitivity make this a viable option for environmental monitoring.

7. Distributed optical fiber sensors

Distributed optical fiber sensing is a technology that provides spatially resolved measurements along a single, continuous optical fiber. This capability allows for continuous, real-time monitoring over long distances, a major advantage over traditional pointwise sensors, even when they are multiplexed. This technology is primarily based on optical scattering phenomena such as Rayleigh, Brillouin, and Raman scattering.

When light propagates through an optical fiber, it interacts with the medium, causing a portion of the light to scatter. This scattering can be elastic, meaning the scattered light retains the same frequency as the incident light, or inelastic, where the frequency is shifted. The scattering spectrum typically shows a central peak at the original incident frequency, which corresponds to Rayleigh scattering. Peaks with a frequency shift are associated with Brillouin and Raman scattering. The characteristics of the scattered light can provide important information about the medium [52].

Rayleigh scattering is an elastic scattering process caused by microscopic inhomogeneities frozen into the fiber during its manufacturing. The scattering intensity is highly dependent on the incident light's wavelength, being stronger for shorter wavelengths [45]. Because its frequency remains the same, Rayleigh backscattering is used in techniques such as optical time-domain reflectometry (OTDR) and optical frequency-domain reflectometry (OFDR) to detect localized changes in physical parameters such as strain or temperature [52].

In turn, Brillouin scattering is an inelastic process resulting from the interaction between light and acoustic phonons. This interaction causes a frequency shift, known as the Brillouin frequency shift (BFS), which is sensitive to both temperature and strain on the fiber [53]. Brillouin-based sensors, such as Brillouin optical time-domain reflectometry (BOTDR) and Brillouin optical time-domain analysis (BOTDA), are examples of techniques that can be employed for long-range monitoring and can simultaneously measure temperature and strain [52].

Additionally, Raman scattering is another inelastic process caused by light interacting with molecular vibrations. Similar to Brillouin scattering, it also results in a frequency shift. The properties of the Raman scattered light are highly dependent on temperature [54]. This makes Raman-based sensors, particularly Raman OTDR (R-OTDR), great candidates for distributed temperature sensing applications [52].

In this context, distributed optical fiber sensors offer several advantages over traditional sensors. Due to the low transmission losses of optical fibers, remote sensing over tens of kilometers is enabled. Similarly to other fiber technologies, these sensors are also immune to electromagnetic interference and can operate in harsh environments, including high temperatures and corrosive substances. Their compact size, flexibility, and light weight make them suitable for applications with limited space.

Thus, this technology is highly valuable for a wide range of applications, especially in the energy sector. For instance, it is used for structural health monitoring of pipelines, bridges, and buildings. These sensors can, for example, detect pipeline leaks by monitoring temperature changes and are utilized for real-time monitoring of power transmission lines for extreme temperatures and strain. In the oil and gas industry, these sensors can monitor wellbore integrity and reservoir performance by detecting temperature gradients. Additionally, distributed sensors are being explored for chemical sensing, such as detecting hydrogen gas concentration and monitoring corrosion [52].

Finally, it is worth mentioning that the area of this family of sensors is a growing field. While the high cost of interrogation systems remains a barrier, ongoing innovations in components and data processing are expected to make these systems more accessible for a broader range of industrial applications.

8. Conclusions

This chapter provided an overview of optical fiber sensors tailored for temperature and humidity monitoring, detailing their fundamental principles and advanced architectures. We've examined distinct sensing platforms, namely fiber Bragg gratings, long-period gratings, Fabry–Perot interferometers, MMI sensors, microstructured optical fibers, and distributed fiber sensors.

Each of these technologies offers promising advantages for optical sensing. FBGs are well-established for their low insertion loss and narrow reflection spectrum, enabling their versatility for many applications, including temperature and humidity sensing when coated with a hygroscopic material. LPGs, by coupling core and cladding modes, demonstrate high sensitivity to external media, a feature often enhanced by polymer coatings such as polyvinyl alcohol or graphene oxide.

In turn, Fabry–Perot interferometers operate based on the principle of optical interference within a cavity. Both intrinsic and extrinsic FPIs can be adapted for temperature sensing by integrating materials with a high thermal expansion coefficient, and for humidity sensing using moisture-sensitive coatings such as chitosan or graphene oxide. MMI sensors, particularly through the SMS structure, allow for exploring the self-imaging phenomenon to detect changes in the surrounding environment, and can be optimized for humidity sensing with specialized coatings.

We also explored microstructured optical fibers, which, with their unique cross-sectional air holes, offer unmatched design flexibility. By infiltrating these holes with temperature-sensitive materials, MOFs can achieve exceptionally high thermal sensitivity. Similarly, coating MOFs with hygroscopic polymers allows for highly sensitive and fast-response humidity monitoring.

Finally, we discussed the topic of distributed optical fiber sensors, which represent a powerful and versatile technology for a variety of monitoring applications. By using the principles of Rayleigh, Brillouin, and Raman scattering, these systems enable continuous, long-range measurement of parameters such as temperature, strain, and vibration. Despite current challenges, like the high cost of interrogation systems, ongoing research in material science, signal processing, and the integration of artificial intelligence promises to enhance the performance and broaden the applicability of these sensors. This will further solidify their role as essential tools for next-generation industrial, environmental, and infrastructure monitoring.

Therefore, this review emphasizes that fiber-based sensors are highly versatile and offer significant advantages for temperature and humidity sensing applications. Their compact design and immunity to electromagnetic interference allow them to operate effectively in harsh environments. The ongoing development regarding microstructure modifications, liquid infiltration techniques, and new functional coatings continues to open new possibilities for creating compact, highly sensitive, and fast-response sensing platforms that support a wide range of emerging sensing technologies.

Acknowledgments

The authors thank FAPEMIG (RED-00046-23, APQ-00197-24, APQ-01401-22, APQ-01618-25, BIS-00254-25), CNPq (305024/2023-0, 402723/2024-4, 409174/2024-6), and Finep for financial support.

Conflict of Interest

The authors declare no conflict of interest.

Abbreviations

FBG	Fiber Bragg grating
LPG	Long-period grating
SMF	Singlemode fiber
MMF	Multimode fiber
PCF	Photonic crystal fiber
HCF	Hollow-core fiber
MOF	Microstructured optical fiber

Author details

Renato Faraco Filho[1], João Gabriel de Carvalho[2], Luiz David Silva[2], Enzo Dantas da Silva[2], Sinei Gonçalves Júnior[2], Flávio Augusto Marques[2], Alexandre Cotta[2], Alexandre Bessa dos Santos[1], Jonas Henrique Osório[2] and Jefferson E. Tsuchida[2]*

1 Faculty of Engineering, Federal University of Juiz de Fora, Juiz de Fora, Brazil

2 Multiuser Laboratory of Optics and Photonics, LaMOF, Department of Physics, Federal University of Lavras, Lavras, Brazil

*Address all correspondence to: jefferson.tsuchida@ufla.br

References

[1] Riza MA, Go YI, Harun SW, Maier R. FBG sensors for environmental and biochemical applications—a review. IEEE Sensors Journal. 2020;**20** (14):7614–7627

[2] Li R, Chen Y, Tan Y, Zhou Z, Li T, Mao J. Sensitivity enhancement of FBG-based strain sensor. Sensors. 2018;**18**(5):1607

[3] Abdulraheem MI, Xiong Y, Zhang W, Chen H, Zhang H, Hu J. Recent applications of fiber Bragg grating sensors in humidity and water content detection in agriculture: A comprehensive review of development, challenges, and future trends. International Journal of Precision Engineering and Manufacturing. 2024;**25**(7):1499–1524

[4] Li T, Guo J, Tan Y, Zhou Z. Recent advances and tendency in fiber Bragg grating-based vibration sensor: A review. IEEE Sensors Journal. 2020;**20** (20):12074–12087

[5] Abro ZA, Hong C, Zhang Y, Siddiqui MQ, Abbasi AMR, Abro Z, Tariq SQB. Development of FBG pressure sensors using FDM technique for monitoring sleeping postures. Sensors and Actuators A: Physical. 2021;**331**:112921

[6] Tosi D. Review of chirped fiber Bragg grating (CFBG) fiber-optic sensors and their applications. Sensors. 2018;**18**(7):2147

[7] Kok SP, Go YI, Wang X, Wong MD. Advances in fiber Bragg grating (FBG) sensing: A review of conventional and new approaches and novel sensing materials in harsh and emerging industrial sensing. IEEE Sensors Journal. 2024;**24**(19):29485–29505

[8] Majumder M, Gangopadhyay TK, Chakraborty AK, Dasgupta K, Bhattacharya D. Fibre Bragg gratings in structural health monitoring—Present status and applications. Sensors and Actuators A: Physical. 2008;**147** (1):150–164

[9] Chen Q, Lu P, Chen Q, Liu P. Fiber Bragg gratings and their applications as temperature and humidity sensors, Atomic. Molecular and Optical Physics. 2008;**6**:235

[10] Oliveira R, Osório JH, Aristilde S, Bilro L, Nogueira RN, Cordeiro CMB. Simultaneous measurement of strain, temperature and refractive index based on multimode interference, fiber tapering and fiber Bragg gratings. Measurement Science and Technology. 2016;**27**:075107

[11] Lins RBE, Aristilde S, Osório JH, Cordeiro CMB, Yanikian CRF, Bicalho AA, Stape THS, Soares CJ, Martins LRM. Biomechanical behaviour of bulk-fill resin composites in class ii restorations. Journal of the Mechanical Behavior of Biomedical Materials. 2019;**98**:255–261

[12] Esposito F, Campopiano S, Iadicicco A. Miniaturized strain-free fiber Bragg grating temperature sensors. IEEE Sensors Journal. 2022;**22**: 16898–16903

[13] Wang X, Sun X, Hu Y, Zeng L, Liu Q, Duan J. Highly-sensitive fiber Bragg grating temperature sensors with metallic coatings. Optik. 2022;**262**:169337

[14] Wang H, Li C, Liang L, Jiang K, Dai S, Wu H, X T. Fast response characteristics of fiber Bragg grating temperature sensors and explosion temperature measurement tests, Sensors and Actuators: A. Physical. 2023;**354**:114236

[15] Zeng F, Tong X, Huang W, Li C, Wang Y, Zou Q. Multipoint temperature measurement system for aero-engine external piping based on arrayed fiber Bragg grating temperature sensors. IEEE Sensors Journal. 2024;**24**: 11102–11112

[16] Aristilde S, Cordeiro CMB, Osório JH. Gasoline quality sensor based on tilted fiber Bragg gratings. Photonics. 2019;**6**:1–8

[17] D'Amato R, Polimadei A, Terranova G, Caponero MA. Humidity sensing by chitosan-coated fibre Bragg gratings (FBG). Sensors. 2021;**21** (10):3348

[18] Tian M, Huang Y, Li C, Lv M. High-performance humidity sensor based on a micro-nano fiber Bragg grating coated with graphene oxide. Optics Express. 2020;**28**(18): 26395–26406

[19] Frazão O, Falate R, Fabris JL, Santos JL, Ferreira LA, Araújo FM. Optical inclinometer based on a single long-period fiber grating combined with a fused taper. Optics Letters. 2006;**31** (20):2960

[20] Stephen RPT, James W. Optical fibre long-period grating sensors: Characteristics and application. Measurement Science and Technology. 2003;**14**(5):49

[21] Shima TSS, Himeno K. A novel temperature-insensitive long-period fiber grating using a boron-codoped-germanosilicate-core fiber, In: Proceedings of Optical Fiber Communication Conference, IEEE, 1997, p. 347, 348

[22] Venugopalan T, Sun T, Grattan K. Long period grating-based humidity sensor for potential structural health monitoring, Sensors and Actuators A. Physical. 2008;**148**(1):57–62

[23] Venugopalan T, Yeo TL, Sun T, Grattan KT. LPG-based PVA coated sensor for relative humidity measurement. IEEE Sensors Journal. 2008;**8**(7):1093–1098

[24] Zheng S. Long-period fiber grating moisture sensor with nano-structured coatings for structural health monitoring. Structural Health Monitoring. 2015;**14**(2):148–157

[25] Tsai Y-T, Wu C-W, Tsai L, Chiang -C-C. Application of graphene oxide-based, long-period fiber grating for sensing relative humidity. Journal of Lightwave Technology. 2020;**39** (12):4124–4130

[26] Zhu C, Zheng H, Ma L, Yao Z, Liu B, Huang J, Rao Y. Advances in fiber-optic extrinsic Fabry–Perot interferometric physical and mechanical sensors: A review. IEEE Sensors Journal. 2019;**23**:6406–6426

[27] Liu J, Jia P, Zhang H, Tian X, Liang H, Hong Y, Liang T, Liu W, Xiong J. Fiber-optic Fabry–Perot pressure sensor based on low-temperature co-fired ceramic technology for high-temperature applications. Applied Optics. 2018;**57**:4211–4215

[28] Li W, Liang T, Jia P, Lei C, Hong Y, Li Y, Yao Z, Liu W, Xiong J. Fiber-optic

Fabry–Perot pressure sensor based on sapphire direct bonding for high-temperature applications. Applied Optics. 2019;**58**:1662–1666

[29] Liu J, Jia P, Feng F, An G, Liang T, Hong Y. MgO single crystals MEMS-based fiber-optic Fabry–Perot pressure sensor for harsh monitoring. IEEE Sensors Journal. 2021;**21**:4272–4279

[30] Chen LH, Li T, Chan CC, Menon R, Balamurali P, Shaillender M, Neu B, Ang XM, Zu P, Wong WC, Leong KC. Chitosan based fiber-optic Fabry–Perot humidity sensor. Sensors and Actuators B: Chemical. 2012;**169**:167–172

[31] Santos JS, Raimundo Jr IM, Cordeiro CMB, Biazoli CR, Gouveia CAJ, Jorge PAS. Characterisation of a nafion film by optical fibre Fabry–Perot interferometry for humidity sensing. Sensors and Actuators B: Chemical. 2014;**196**:99–105

[32] Li C, Yu X, Zhou W, Cui Y, Liu J, Fan S. Ultrafast miniature fiber-tip Fabry–Perot humidity sensor with thin graphene oxide diaphragm. Optics Letters. 2018;**43**:4719–4722

[33] Bian C, Hu M, Wang R, Gang T, Tong R, Zhang L, Guo T, Liu X, Qiao X. Optical fiber humidity sensor based on the direct response of the polyimide film. Applied Optics. 2018;**57**:356–361

[34] Wu Q, Qu Y, Yuan J, Wan S, Wu T, He X, Liu B, Liu D, Ma Y, Semenova Y, Wang P, Xin X, Farrell G. Singlemode-multimode-singlemode fiber structures for sensing review. IEEE Sensors Journal. 2021;**21**:12734–12751

[35] Osório JH, Guimarães WM, Peng L, Franco MAR, Warren-Smith SC, Ebendorff-Heidepriem H, Cordeiro CMBMB. Exposed-core fiber multimode interference sensor. Results in Optics. 2021;**5**:100125

[36] Silva S, Pachon EEGP, Franco MMAR, Hayashi JG, Malcata FX, Frazão O, Jorge P, Cordeiro CCMB. Ultrahigh-sensitivity temperature fiber sensor based on multimode interference. Applied Optics. 2012;**51**:3236–3242

[37] Wang P, Ni K, Wang B, Ma Q, Tian W. Methylcellulose coated humidity sensor based on Michelson interferometer with thin-core fiber. Sensors and Actuators A: Physical. 2019;**288**:75–78

[38] Ma QF, Tou ZQ, Ni K, Lim YY, Lin YF, WYR, Zhou MH, Shi FF, Niu L, Dong XY, Chan CC. Carbon-nanotube / polyvinyl alcohol coated thin core fiber sensor for humidity measurement. Sensors and Actuators B: Chemical. 2018;**257**:800–806

[39] Wang X, Farrell G, Lewis E, Tian K, Yuan L, Wang P. A humidity sensor based on a singlemode-side polished multimode–singlemode optical fibre structure coated with gelatin. Journal of Lightwave Technology. 2017;**35**:4087–4094

[40] Urrutia E, Ayesta I, Azkune M, Illarramendi MA, Zubia J. Recent advances in sensor applications of microstructured optical fibers: A review. IEEE Sensors Journal. 2025;**25**:18584–18607. DOI: 10.1109/JSEN.2025.3558308

[41] Gangwar RK, Kumari S, Pathak AK, Gutlapalli SD, Meena MC. Optical fiber based temperature sensors: A review. Optics. 2023;**4**:171–197

[42] Qian W, Zhao C-L, He S, Dong X, Zhang S, Zhang Z, Jin S, Guo J, Wei H. High-sensitivity temperature sensor based on an alcohol-filled photonic crystal fiber loop mirror. Optics Letters. 2011;**36**:1548

[43] Reyes-Vera E, Cordeiro C, Torres P. Highly sensitive temperature sensor using a Sagnac loop interferometer based on a side-hole photonic crystal fiber filled with metal. Applied Optics. 2017;**56**:156

[44] Chesini G, Osório J, Serrao V, Franco M, Cordeiro C. Metal-filled embedded-core capillary fibers as highly sensitive temperature sensors. IEEE Sensors Letters. 2018;**2**:1–4

[45] Osório JH, Amrani F, Delahaye F, Dhaybi A, Vasko K, Melli F, Giovanardi F, Vandembroucq D, Tessier G, Vincetti L, Debord B, Gérôme F, Benabid F. Hollow-core fibers with reduced surface roughness and ultralow loss in the short-wavelength range. Nature Communications. 2023;**14**:1146. DOI: 10.1038/s41467-023-36785-6

[46] Li Z, Wu Z, Li Z, Ou L, Zhang W, Lai Z, Zhang Y, Xie M, Tang J, Zhu W, Zheng H, Zhong Y, Deng X, Zou X, Chen Z, Yu J. High-sensitivity and fast-response fiber optic temperature sensor using an anti-resonant reflecting optical waveguide mechanism. Photonics Research. 2023;**11**:1397. DOI: 10.1364/PRJ.492840

[47] Niu P, Liu S, Cheng W, Li L. Liquid modified anti-resonant reflecting guidance in silica capillary for sensing applications. IEEE Access. 2022;**10**:53676–53681. DOI: 10.1109/ACCESS.2022.3176458

[48] Rodrigues GL, Cordeiro CMB, Amrani F, Gérôme F, Benabid F, Osório JH. Temperature sensing with a liquid-filled hollow-core photonic crystal fiber. In: 2023 International Conference on Optical MEMS and Nanophotonics (OMN) and SBFoton International Optics and Photonics Conference (SBFoton IOPC), IEEE, 2023, p. 1–2. 10.1109/OMN/SBFotonIOPC58971.2023.10230983

[49] Junior SFG, Marques FAM, Cotta AAC, Tsuchida JE, Dos Santos AB, Osório JH. Design of a temperature sensor based on partially filled hollow-core optical fibers. In: SBFoton, IEEE, 2024, p. 1–3. 10.1109/SBFotonIOPC62248.2024.10813524

[50] Shrivastav AM, Gunawardena DS, Liu Z, Tam H. Microstructured optical fiber based Fabry–Pérot interferometer as a humidity sensor utilizing chitosan polymeric matrix for breath monitoring. Scientific Reports. 2020;**10**:6002

[51] Li J, Tong Z, Jing L, Zhang W, Qin J, Liu J. Fiber temperature and humidity sensor based on photonic crystal fiber coated with graphene oxide. Optics Communications. 2020;**467**:125707

[52] Lu P, Lalam N, Badar M, Liu B, Chorpening BT, Buric MP, Ohodnicki PR. Distributed optical fiber sensing: Review and perspective. Applied Physics Reviews. 2019;**6**:041302

[53] Xiao W, Zhang Z, Qin Y, Wang C, Xiong C, Osório JH, Jiang W, Shen Y, Xiao L. Discriminative sensing of acoustic impedance and temperature based on spatial-division multiplexed forward Brillouin scattering in multicore fiber. IEEE Sensors Journal. 2025;**25**:21633–21641

[54] Liu Y, Ma L, Yang C, Tong W, He Z. Long-range Raman distributed temperature sensor with high spatial and temperature resolution using graded-index few-mode fiber. Optics Express. 2018;**26**:20562–20571

Chapter 3

Temperature and Humidity Sensors' Application in Power Equipment

Lifan Yang

Abstract

In the evolving landscape of smart grids, power equipment – ranging from switchgear to transformers, cables, and generators – requires robust condition monitoring to ensure uninterrupted energy delivery. A variety of sensors are used in monitoring of these important power equipment. Temperature and humidity sensors play a pivotal role in this monitoring, as overheating is a primary indicator of equipment degradation or failure. However, power equipment operates in diverse harsh environments: transformers endure oil degradation and thermal cycles, cables face soil thermal resistance and load fluctuations, and switchgear encounters electromagnetic interference and short-circuit stresses. This chapter describe the different ways of using temperature and humidity sensors in power equipment domain and analyze the lifespan of these sensors.

Keywords: power equipment, temperature sensor, humidity sensor, condition monitoring, transformer, cable joint, switchgear

1. Introduction

Condition monitoring of critical power equipment across multiple dimensions – encompassing temperature, humidity, partial discharge, gas pressure, contact wear, arc flash, and power quality – proves highly valuable in addressing equipment malfunctions during operation. Moreover, dynamic trend analysis derived from monitoring data offers more scientific guidance for the planning of preventive maintenance strategies [1].

In real-time condition monitoring, thermal measurement is particularly common and critical. After years of development, online temperature monitoring technologies have evolved into a diverse range of approaches [2], including passive wireless temperature measurement (e.g., surface acoustic wave-based sensing), passive wired temperature measurement (e.g., optical fiber sensing), active wireless temperature measurement (e.g., inductively powered wireless sensing), and infrared temperature measurement. These methods have achieved reliable temperature monitoring across different operational environments, with varying degrees of precision and adaptability to harsh conditions.

Ambient humidity measurement and regulation are equally essential, especially for power equipment or enclosed electrical components that are sensitive to moisture levels. Humidity monitoring techniques have diversified into several specialized approaches, including capacitive humidity sensing (e.g., polymer-based capacitive sensors), resistive humidity sensing (e.g., ceramic resistive elements), optical humidity measurement (e.g., fiber-optic humidity transducers), and wireless humidity sensing (e.g., battery-powered IoT-enabled sensors). These methods have achieved humidity monitoring capabilities to varying extents.

However, both temperature and humidity measurement technologies face distinct challenges in practical applications on power equipment. For temperature monitoring, issues often arise from harsh operational environments: high electromagnetic interference in switchgear can disrupt wireless signal transmission, especially for passive wireless systems, leading to data loss or inaccuracies. Additionally, surface-mounted sensors may suffer from poor thermal contact with high-voltage components, causing delayed or distorted readings, while optical fiber sensors, though resistant to EMI, are vulnerable to mechanical damage during equipment maintenance.

Humidity measurement, on the other hand, struggles with environmental constraints such as condensation and dust accumulation. Enclosed electrical cabinets, where humidity levels need strict control, can trap moisture that adheres to sensor surfaces, gradually degrading the sensitivity of capacitive or resistive elements. Moreover, rapid temperature fluctuations within equipment – common during load changes – can induce temporary humidity measurement errors due to thermal effects on sensing materials.

Both technologies also face challenges in long-term stability: temperature sensors may drift due to prolonged exposure to high temperatures, while humidity sensors require frequent calibration to counteract drift caused by chemical aging, especially in polluted or corrosive atmospheres typical in industrial power settings.

2. Temperature monitoring technologies for power equipment

There are all kinds of electrical equipment that need to be monitored online. Generally speaking, the higher the voltage level of the equipment and the larger its capacity, the greater its importance, and the more complex and advanced the monitoring required [3].

2.1 Switchgear temperature sensing

Temperature measurement methods for power switchgear (such as circuit breakers, disconnectors, switchgears, etc.) need to target heat-prone components like contacts, busbar connections, and infeed terminations. Considering their characteristics of high voltage, enclosed structures, or strong electromagnetic environments, safe, accurate, and practical methods are selected.

2.1.1 Contact temperature measurement

Temperature is obtained through physical contact between sensors and measured points where heat-generating components are present. Direct contact with the conductor ensures high accuracy [4], while insulation treatment is required.

2.1.1.1 Thermistor temperature measurement

Platinum resistors (Pt100) are affixed or embedded in contacts, connection points, and other critical components of switchgear. Leveraging their inherent property of resistance variation with temperature – where resistance increases as temperature rises – temperature signals are transmitted to secondary instruments via wires. These sensors offer high accuracy (±0.1°C–0.5°C) and rapid response times (1–5 seconds). However, they require direct physical contact with the measured points, and in high-voltage environments, insulation must be reinforced – for instance, by using high-temperature-resistant insulated wires.

2.1.1.2 Thermocouple temperature measurement

Thermocouple probes, which operate based on the "Seebeck effect" of two dissimilar metals, are welded to the surface of heat-generating components. The thermoelectric potential induced by temperature differences is transmitted to measuring instruments via compensation wires. These sensors boast an extensive temperature range (-200°C–1300°C), making them ideal for high-temperature components – such as circuit breaker contacts, which can exceed 800°C during short-term overheating events. While thermocouples offer cost-effectiveness, their accuracy is moderate (±1°C – ±3°C), and they require cold junction temperature compensation to ensure measurement reliability.

2.1.1.3 Optical fiber temperature measurement

Fiber Bragg Grating (FBG) temperature measurement works by attaching fiber Bragg grating sensors to the measured points. This technology boasts key features such as full insulation (as fibers are non-conductive) and strong resistance to electromagnetic interference, making it highly suitable for high-voltage switchgears and gas-insulated switchgear (GIS) equipment. It delivers high accuracy within the range of ±0.5°C–±1°C, offers temperature resistance up to 200°C, and supports long signal transmission distances.

Another method is distributed fiber optic temperature sensing (DTS) which could be laid along switchgear busbars.

2.1.1.4 Integrated digital sensors

Semiconductor digital temperature sensors are directly installed at the measured points, outputting temperature data via standardized protocols. With a compact chip-type design, these sensors are highly suitable for installation in narrow spaces within switchgears (**Figure 1**). However, they have a limitation in temperature resistance, typically operating within a range of -55–125°C, which restricts their application in high-temperature scenarios.

Digital sensors have built-in energy harvesting modules using different principles. An electromagnetic induction coil that obtains electricity from the leakage magnetic field of high-voltage equipment is the most common method. Some sensors have replaceable lithium batteries with a 3- to 10-year lifespan. Some sensors harvest energy by using wireless radio frequency technologies, e.g., Surface Acoustic Wave (SAW) and RFID. For outdoor switchgear (e.g., pole-mounted circuit breakers), a solar auxiliary power supply is also used.

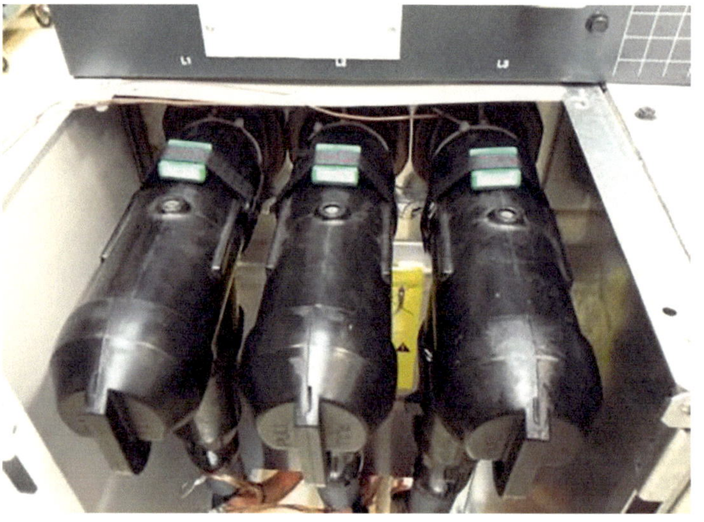

Figure 1.
Bundling temperature measurement scheme for Ring Main Unit (A common equipment in electrical distribution network).

Aiming at the difficulty of wiring in high-voltage switchgear (e.g., GIS, circuit breakers), many digital sensor models are equipped with built-in wireless communication modules (such as ZigBee, LoRa, NB-IoT, or Bluetooth), enabling cable-free data transmission that eliminates wiring constraints in confined or hard-to-reach areas, greatly enhancing installation flexibility. Their digital signal transmission ensures strong anti-interference capability, effectively resisting electromagnetic interference in complex electrical environments, while also supporting flexible multi-point networking for centralized monitoring of multiple measuring points (**Figure 2**).

2.1.2 Non-contact temperature measurement

Non-contact temperature measurement relies on optical, infrared, and other remote-sensing technologies to detect temperature without physical contact,

Figure 2.
Circuit breaker contact arm embedded temperature sensor.

effectively eliminating the risk of high-voltage electric shock. This approach is particularly well-suited for high-voltage or enclosed environments, where direct access to measurement points is restricted or hazardous, ensuring safe and reliable temperature monitoring.

2.1.2.1 Infrared temperature measurement

Include point-type infrared sensors and infrared thermal imagers, both leveraging infrared radiation detection for temperature monitoring.

Point-type infrared sensors operate by focusing infrared radiation from a specific measured point through a lens, where infrared detectors convert the radiant energy into electrical signals to calculate temperature. They support both fixed installation inside the observation window of switchgears and portable use by maintenance engineers. Fast response times (<1 s) and an accuracy range of ±2°C–±5°C make them well-suited for monitoring single hotspots like circuit breaker contacts.

In contrast, infrared thermal imagers detect infrared radiation across an entire measured area using an infrared focal plane array (FPA), generating a temperature field distribution map (thermal image) that intuitively visualizes the overall temperature profile, enabling simultaneous monitoring of multiple points (such as all connections within switchgears) to quickly pinpoint local overheating, with higher accuracy (±1°C or ±1% of the reading) (**Figure 3**).

2.1.3 Traditional auxiliary temperature measurement methods

2.1.3.1 Temperature-indicating wax tablets/stickers

Temperature-indicating materials, which change color at specific temperatures, are affixed to the surface of switch connections to intuitively identify overheating. Their appeal lies in extremely low cost and easy installation, making them suitable for scenarios where only qualitative assessments are needed – though they cannot provide precise temperature readings and require manual inspection.

2.1.3.2 Handheld infrared thermometers

During manual inspections, handheld infrared thermometers are used to measure temperatures on the surface of switchgear (e.g., cabinet, busbar joints). Flexible and portable, suitable for regular inspections, but significantly affected by human operation (e.g., distance, angle) and unable to perform real-time monitoring.

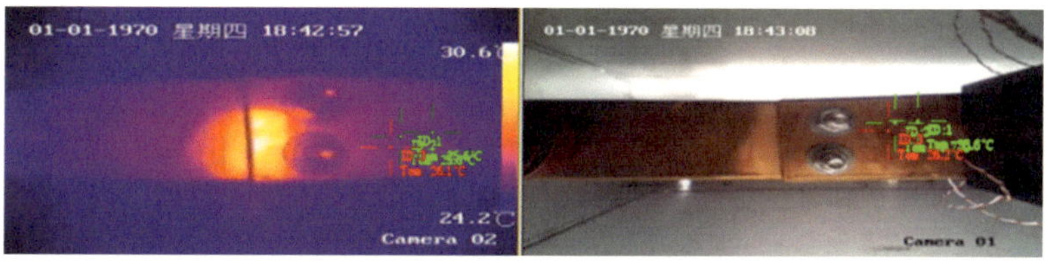

Figure 3.
The thermal imaging (left) and visible light imaging (right) of busway inside switch cabinet.

2.1.4 Summary

Temperature measurement for electrical switchgear should be selected based on voltage level, environmental characteristics, and monitoring requirements:

- low-voltage equipment – thermistors and integrated sensors are preferred (low cost, medium accuracy);
- high-voltage enclosed equipment – FBG and passive wireless temperature sensors are preferred (anti-electromagnetic interference, safety);
- outdoor or large-scale monitoring – active wireless and infrared thermal imagers are suitable (flexible, visual);
- simple inspections – temperature-indicating wax tablets and handheld infrared thermometers can be used (low cost, convenient).

In practical applications, a combination of "online monitoring + regular inspection" is often adopted to timely detect overheating faults (e.g., poor contact, insulation aging) and avoid equipment damage or power outages.

2.2 Transformer temperature monitoring

2.2.1 Temperature measurement methods for oil-immersed transformers

Oil-immersed transformers use insulating oil as the cooling and insulating medium. The core of temperature measurement is to monitor the oil temperature and winding temperature, which best reflect the transformer load and heating status.

2.2.1.1 Oil Temperature Measurement

Pressure-type thermometer: Consists of a temperature bulb (inserted into the oil), a capillary tube, and a gauge head. The liquid (such as alcohol, kerosene) in the temperature bulb expands when heated to generate pressure, which is transmitted to the gauge head through the capillary tube to drive the pointer, with moderate accuracy (±1°C–± 3°C). It can provide remote indication (the capillary tube can be up to meters long). Some models have a contact output for an over-temperature alarm (**Figure 4**).

This type of temperature measurement is also used in other oil-immersed equipment, such as large electric reactors and rectifier transformers [5].

Thermistor thermometer: A platinum resistance (Pt100) sensor is inserted into the top of the oil tank. It utilizes the characteristic that its resistance changes with temperature. The temperature is measured with high accuracy (±0.1–±0.3°C) and transmitted remotely through secondary instruments (such as digital displays, PLCs), supporting digital signal output, and can be connected to monitoring systems.

2.2.1.2 Winding temperature measurement

The temperature of transformer windings is usually higher than the oil temperature since the windings are the main heat source. Experimental results show

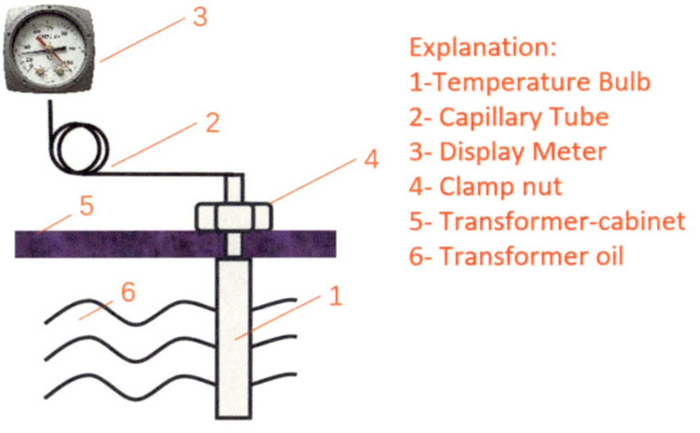

Figure 4.
A typical pressure-type temperature measuring system for transformer oil.

that transformer winding temperature is more prominently correlated with the operating state of the transformer than oil temperature [6].

Winding temperature indicator (WTI): Based on oil temperature measurement, it collects the winding current through a current transformer and uses the "current thermal effect" to simulate the additional temperature rise caused by winding loss (i.e., "oil temperature + current compensation temperature rise = winding temperature"). Composed of an oil temperature sensor, a current transformer, and a compensation heating element. Obviously, it is an indirect measurement; no need to directly contact the windings to avoid insulation damage, moderate cost, and accuracy (±3℃ ~ ±5℃) is affected by the precision of compensation.

Fiber Bragg grating temperature measurement: Fiber Bragg grating (FBG) sensors are implanted inside windings to enable direct measurement of winding temperatures. This technology features strong anti-electromagnetic interference capabilities – making it ideal for high-voltage environments – and high accuracy (±0.5℃). However, it entails higher costs and a more complex installation and calibration process, which is why it is primarily utilized in large transformers or smart grid renovation projects (**Figure 5**).

2.2.2 Temperature Measurement Methods for Dry-type Transformers

Dry-type transformer windings are cast in insulating materials, and the core of temperature measurement is the winding temperature.

Pre-embedded Thermistor: During the modeling of transformer windings, thermistors (e.g. Pt100) are pre-embedded in the hot spot positions of the windings (such as the middle of the windings) to directly measure the winding temperature with fast response and high accuracy. The signal can be connected to a temperature controller to realize over-temperature alarm or start cooling fans.

In addition to Pt100 sensors, negative temperature coefficient (NTC) thermistors are also utilized in certain scenarios, favored for their low cost and compact size. However, they suffer from relatively poor accuracy, typically within a ±5℃ range. As such, their application is largely confined to small transformers (below 100kVA) where only basic temperature control is required

Externally attached Temperature Measurement: Self-powered wireless sensors are assembled on medium-voltage (MV) and low-voltage (LV) junction connections to enable early detection of hot-spot at these critical joints (**Figure 6**). This proactive

Figure 5.
Multi-point temperature sensors installation on transformer winding.

monitoring capability helps prevent unplanned downtime associated with overheating connections—while also facilitating real-time tracking of thermal behavior changes in transformers.

2.2.3 Summary

Temperature measurement for power transformers needs to consider safety (avoiding damage to insulation), accuracy (reflecting the real heating status), and practicality (facilitating integrated monitoring). In traditional methods, oil-immersed transformers mainly use "oil temperature + winding temperature indicator," and dry-type transformers mainly use "pre-embedded Pt100". New

(a) on low-voltage connections (b) on medium-voltage connections and tapping links

Figure 6.
External temperature sensors attached on dry-type transformer.

intelligent temperature measurement technologies (optical fiber, wireless, infrared thermal imaging) are more suitable for online monitoring needs of digital and intelligent power grids. In practical applications, a multiple temperature measurement scheme of "local indication + remote monitoring + abnormal alarm" is often adopted to ensure the safe operation of transformers.

2.3 Cable and overhead line thermal monitoring

As core carriers of power transmission, underground cables and overhead lines require online temperature monitoring solutions tailored to their structural characteristics (cables are mostly enclosed, while overhead lines are exposed outdoors) and operating environments (high voltage, strong electromagnetic fields, harsh weather, etc.).

2.3.1 Temperature monitoring methods for cables

Cables, especially high-voltage ones, are often buried underground, routed through pipes, or laid in cable trays, with enclosed environments prone to heat accumulation. Key monitoring points include cable joints (intermediate joints, terminations) and densely laid sections, which would cause overheating or even fire.

2.3.1.1 Distributed fiber optic sensing

A dedicated temperature-sensing fiber is laid along the entire length of the cable (e.g., tightly wrapped around the cable insulation or laid alongside the cable). It real-time monitors the temperature distribution along the entire fiber, with positioning accuracy up to the meter level. These features enable continuous temperature measurement of the entire cable rather than single-point monitoring, quickly identifying localized overheating segments (e.g., poor joint contact, insulation aging).

Resistant to electromagnetic interference (fibers are non-conductive), wide temperature range (-50°C–300°C), and long lifespan (matching the cable's service life), making it suitable for high-voltage cables (110kV and above). While it delivers high measurement accuracy of ±1°C, it comes with relatively higher costs due to the need for specialized demodulation equipment.

2.3.1.2 Wireless sensor node monitoring

Small wireless temperature sensors (with built-in NTC or Pt100) are installed at hotspots such as cable joints and terminals. Sensors are fixed via magnetic attraction or adhesion, transmitting temperature data to a gateway using wireless protocols like ZigBee, LoRa, or NB-IoT. Flexible installation (no wiring required) makes it suitable for monitoring local hotspots like cable joints, with a lower cost than DTS.

This monitoring device is powered usually by two modes: battery-powered (with a 3 to 5-year lifespan) or electromagnetic induction (harvesting energy from the cable's magnetic field to achieve maintenance-free operation), enabling it to adapt well to enclosed environments such as cable wells and trays. It offers moderate measurement accuracy of ±2°C–±3°C, though its application is limited by data transmission distance, requiring it to stay within signal coverage (**Figure** 7).

2.3.1.3 Contact wired temperature monitoring

Pt100 thermistors or thermocouples are directly affixed to or embedded in cable surfaces or joints, with temperature signals transmitted via shielded cables to monitoring devices. This configuration delivers high measurement accuracy

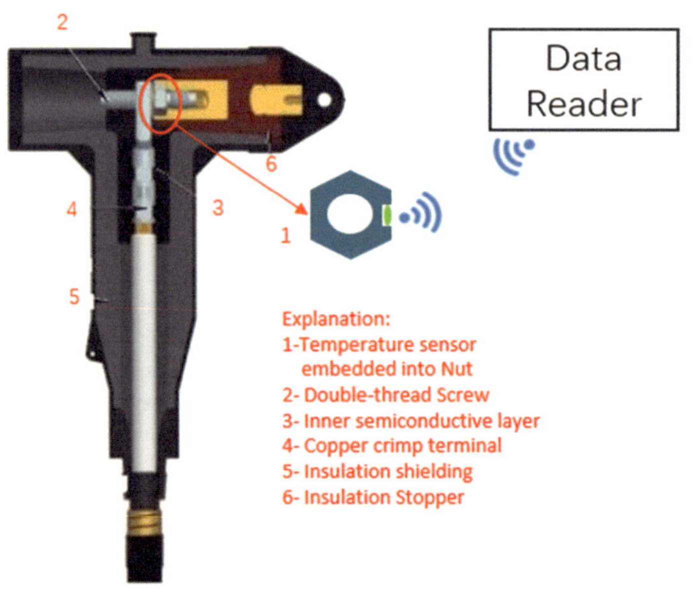

Figure 7.
A temperature sensor embedded inside T-type cable connector.

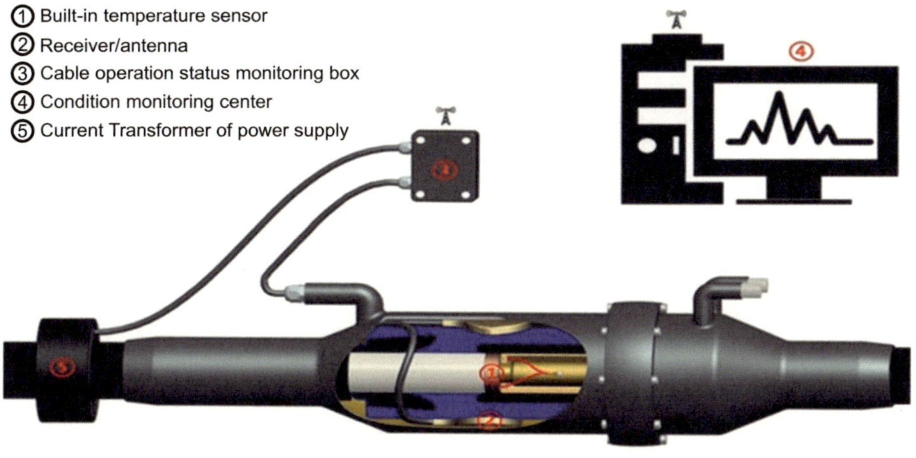

Figure 8.
A temperature sensor embedded inside cable joint.

(±0.5°C–±1°C) and rapid response times. However, it requires dedicated wiring and is inherently susceptible to electromagnetic interference – necessitating robust shielding measures. These constraints, coupled with high wiring costs, limit its practical application in long-distance cable monitoring scenarios (**Figure 8**).

2.3.2 Temperature monitoring methods for overhead lines

Overhead lines are exposed outdoors, operate at high voltages (110 kV and above), span long distances (up to dozens of kilometers), and are affected by wind, ice, and sunlight. Key monitoring points include joints (tension clamps, splice sleeves) and lowest sag points (prone to overheating). Monitoring methods must meet requirements for non-contact operation, resistance to harsh environments, and safe insulation.

2.3.2.1 Infrared thermography monitoring

Infrared thermal imagers are fixedly installed on poles or adjacent structures. Leveraging optical lenses, they capture infrared radiation emitted by overhead lines and joints from a distance (ranging 10–50 m). Moreover, it allows for simultaneous monitoring of multi-phase conductors and joints, offering high visualizability – thermal images intuitively present temperature distribution, enabling quick location of overheating points (such as localized high temperatures caused by loose joints) with an accuracy of ±1°C– ±2°C. However, they are affected by weather (rain, snow, or fog), and regular lens cleaning is required to maintain optimal operation.

In addition to fixed infrared thermal imagers, drones equipped with infrared thermal imaging modules are increasingly used for overhead line inspection. Drones can maneuver flexibly around power lines, reaching areas that fixed installations might miss or struggle to view clearly, thereby serving as a valuable complement to fixed imager systems (**Figure 9**).

In some cases, infrared temperature measurement is combined with UV sensor which detects corona discharge (a phenomenon often linked to overheating), to comprehensively assess insulation defects or poor contact issues in overhead lines.

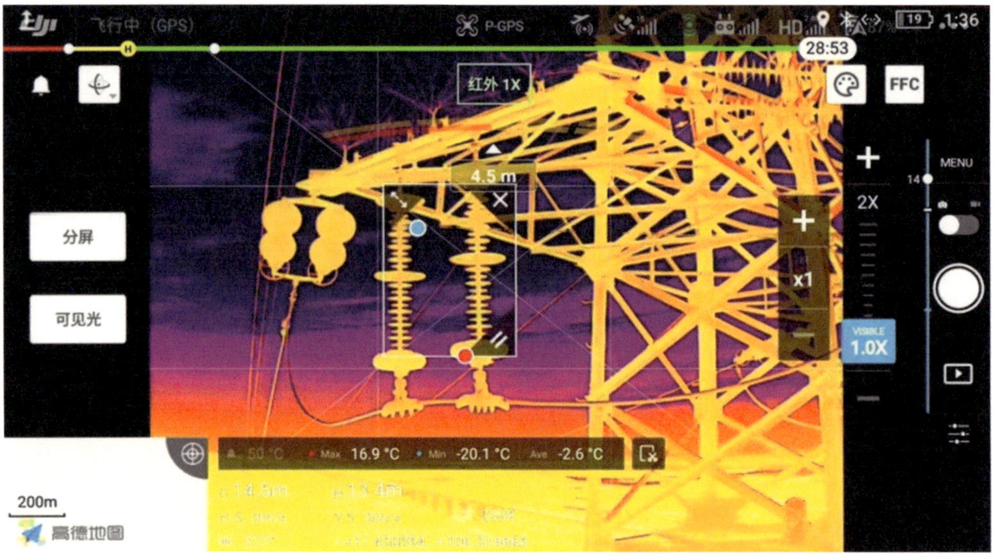

Figure 9.
Infrared thermal imagery for transmission line by patrolling drone.

Going beyond temperature monitoring, it correlates thermal data with discharge phenomena, thereby enhancing the accuracy of fault warnings. This integrated approach is particularly well-suited for condition assessment of aging power lines.

2.3.2.2 Wireless passive temperature sensors

Sensors are installed on overhead line joints (e.g., tension clamps), where they make direct contact with conductors to ensure high measurement accuracy (±1°C–±2°C). Featuring a passive design, these sensors harvest energy from conductor currents via electromagnetic induction, eliminating the need for batteries. They transmit temperature data to pole-mounted receivers using wireless communication technologies such as RFID, LoRa, or NB-IoT.

Designed to suit high-voltage environments, they are maintenance-free with a lifespan matching that of the line, while boasting resistance to harsh outdoor conditions, including high temperatures, water, and ultraviolet radiation (**Figure 10**).

2.3.2.3 Fiber Bragg grating sensors

Fiber Bragg grating (FBG) sensors are fixed to the overhead line surface or joints via specialized fixtures. Given the sensitivity of FBG sensor to both temperature and strain, sensors are encapsulated using a stress-free packaging combined with fixtures (i.e., force-free encapsulation) to mitigate the issue of temperature-strain cross-sensitivity [7].

2.3.3 Summary

Cable temperature monitoring primarily relies on contact and distributed methods (e.g., DTS, wireless nodes), focusing on precise monitoring of entire lines

Figure 10.
Temperature sensor installed on clip of transmission line.

or local hotspots. Overhead line monitoring emphasizes non-contact and long-distance methods (e.g., infrared thermography, wireless passive sensors), prioritizing safety and resistance to harsh environments. In practice, multiple methods are often combined (e.g., DTS + infrared thermography) to achieve "overall monitoring + targeted tracing," ensuring safe operation of power lines.

3. Discussion on the lifespan of sensors

Regardless of the types of temperature measurement scheme employed, the lifespan of sensors remains a primary concern raised by power equipment operators and maintenance engineers. A key industry focus has long been whether sensors can achieve a lifespan comparable to that of the power equipment they monitor.

These temperature measurement systems, composed of electronic components, are inherently more vulnerable and typically have shorter operational lifespans than the power equipment itself. Merely integrating them through simple installation methods, without rigorous engineering consideration, often results in high failure rates and reduced service life for the temperature sensors, and even poses latent risks to the normal operation of the power equipment being monitored.

3.1 Core factors affecting the lifespan of temperature sensors

The operating environment of power equipment (such as transformers, switchgears, cable joints, and transmission lines) is often characterized by high temperature, high humidity, strong electromagnetic interference, vibration, dust, or corrosive gases. These factors directly affect the service life of sensors.

3.1.1 Environmental erosion

3.1.1.1 High temperature and temperature fluctuations

During operation, the internal temperature of power equipment may remain between 50℃ and 90℃ for a long time (or even higher; for example, the hot-spot temperature of transformer windings can exceed 140℃). Sensitive components of sensors (such as thermoelectric materials in thermocouples, resistance wires in thermal resistors, and coating layers of optical fibers) are prone to material aging and measurement drift under continuous high temperatures or frequent cold-heat cycles, leading to shortened lifespan.

3.1.1.2 Humidity and corrosive environments

In scenarios such as outdoor equipment or underground cable trenches, high humidity, condensation, or environments containing corrosive gases (such as sulfur and chlorine) can accelerate the oxidation and cracking of sensor packaging materials (such as housings and cable insulation layers) and may even cause internal circuit short circuits or poor contact.

3.1.2 Mechanical and electrical stress damage

3.1.2.1 Mechanical stress damage

Vibration during the operation of equipment like transformers and transmission line galloping, or improper fastening during sensor installation (such as mechanical stress caused by over-tightening), may lead to breakage of sensor leads, detachment of solder joints, or structural damage to sensitive components (such as fiber Bragg gratings).

3.1.2.2 Electrical stress damage

Strong electromagnetic fields around high-voltage equipment may interfere with the signal transmission circuits of sensors (such as analog-output thermocouples and thermal resistors), especially significantly affecting the communication modules of wireless transmission sensors. Long-term electromagnetic impact can cause aging and failure of components such as chips and capacitors. Meanwhile, intense electrical surges could “kill” these sensors instantly.

3.1.3 Material and process defects

The material stability of core sensor components (such as the purity of platinum wires in platinum resistors, and the fatigue resistance of quartz materials in optical fibers) and packaging processes (such as sealing performance and quality of anti-corrosion coatings) directly determine the basic lifespan. Inferior materials or process defects can cause performance degradation, such as reduced accuracy and delayed response, in a short period.

3.1.4 Lack of maintenance

Dust and oil accumulated on the sensor surface can affect the temperature measurement accuracy of infrared sensors; long-term lack of cleaning may lead to false alarms. Oxidation or looseness of terminal blocks in wired sensors, if not handled in time, can cause increased contact resistance, indirectly accelerating the aging of sensor circuits due to excessive load. Failure to replace the depleted battery in a timely manner could cause the sensor to not function at all.

3.2 Key measures to extend the lifespan of temperature sensors

The lifespan of sensors can be extended through the following methods:

3.2.1 Proper selection

Selecting the proper type of sensor is of the first importance. The following criteria can be used for selecting sensors:

- Temperature range and accuracy
- Chemical resistance
- Wear resistance and vibration resistance
- Installation limitation (may need to match existing equipment; existing aperture can determine the probe diameter)

For example, thermocouples of letter-designated types (R, S, B, J, T, E, K, N, C, and A) have their reference functions and tolerance specifications clearly [8]. When measuring the temperature of transformer windings, K-type or S-type thermocouples are recommended as the primary choices in the high-temperature-resistant scenario; for high-humidity/corrosive environments, sensors should be selected with a minimum protection rating of IP67 and corrosion-resistant housings; for strong electromagnetic environments, optical fiber sensors or wired sensors integrated with electromagnetic shielding designs are prioritized to ensure measurement stability.

It is advisable to prioritize maintenance-free solutions, such as wireless sensors utilizing electromagnetic induction for power harvesting, which eliminate the need for battery replacement. Core components should be constructed from industrial-grade materials, such as high-precision platinum resistors and fatigue-resistant optical fibers, while packaging processes must comply with relevant power industry standards, such as those established by UL and IEC, to guarantee long-term reliability.

3.2.2 Optimize installation and protection design

During installation, avoid direct contact between sensors and vibration sources, and fix them with shock-absorbing brackets; use gold-plated or silver-plated terminal blocks to reduce oxidation; add rainproof and anti-corrosion shields for outdoor sensors, and equip sensors in cable trenches with dehumidification devices to reduce humidity impact.

3.2.3 Strengthen maintenance and monitoring

Regularly clean the lenses of infrared sensors and check the tightness of wired sensor connections; periodically calibrate the battery voltage of wireless sensors and replace low-power batteries in a timely manner; use the "health status monitoring" function of sensors themselves (such as self-diagnostic output of some intelligent sensors) to early warn of performance degradation.

3.3 Summary

The lifespan of temperature sensors in power equipment is the result of the combined effect of "environmental adaptability + material technology + maintenance management". Generally, it is necessary to match the sensor lifespan with the equipment life cycle (for example, the design life of transformers is 20–30 years). Through scientific selection, standardized installation, and regular maintenance, the sensor lifespan can be extended to 8–15 years. This ensures long-term reliability of temperature monitoring for power equipment and reduces the risk of failures.

Author details

Lifan Yang
Schneider Electric, Shanghai, China

*Address all correspondence to: lifan.yang@se.com

References

[1] Guide for maintenance of distribution systems, Schneider Electric. 2020

[2] Chenjiong L, Zhen L, Yalin W. Summary of research on online monitoring method of cable joint temperature. Electrical Engineering. 2019;**20**(5):1–4, 9

[3] Yang W, Xiaohua W, Mingzhe R, and Shenli J. Online temperature measurements with infrared technology on high voltage device. Proceedings of the CSEE. 2002;**22**(9):113–117

[4] Lifan Y, Analysis of the Lifespan of Temperature Sensors in Medium-Voltage Switchgear, 2024 China International Conference on Electricity Distribution (CICED), September 12–13, 2024

[5] DL/T 1400 -2015, Calibration specification for local temperature measuring equipment of oil transformer

[6] Wang En etc.. Multi Point Temperature Monitoring of Oil Immersed Transformer Based on Fiber Bragg Grating. High voltage engineering. 2017;**43**(5):1543–1549 [Accessed: 2017-May-31]

[7] Yunshui X. Temperature Monitoring of Transmission Conductor Based on Optical Fiber Sensing. Yunnan Electric Power. 2018;**46**(4)

[8] IEC 60584-1. 2013 Thermocouples - Part 1: EMF specifications and tolerances

Chapter 4

Seasonal Variations in Temperature-Humidity Index (THI) and Their Effects on Physiological, Reproductive, and Productivity Performance of Rabbit Does Raised in Benin City, Nigeria

O.D. Asemota, O.E. Okhale and A.M. Orheruata

Abstract

This study evaluated the effects of seasonal temperature-humidity index (THI) variations on the physiological responses, reproductive traits, and productivity of rabbit does in Benin City, Nigeria. Sixty (60) composite rabbits (50 does, 10 bucks) were monitored for 12 months (January–December 2013) under natural conditions. Daily ambient temperature and relative humidity were used to calculate THI, while rectal temperature (RT), respiratory rate (RR), pulse rate (PR), conception rate (CR), gestation length (GLT), litter size (LTS), pre-weaning mortality (PWMor), and kit growth were recorded. THI ranged from 29.0 ± 0.19 in the late rainy season to 32.0 ± 0.14 in the late dry season, surpassing the thermal comfort threshold (THI >30) during the latter. Elevated THI ($P < 0.05$) increased physiological stress indicators and reduced reproductive performance, with conception rates dropping to 20%, gestation extending to 32 ± 0.23 days, and PWMor rising to 73.1%. Lower THI improved reproductive outcomes. Early parities were more heat-sensitive. Climate-smart housing, breeding in cooler months, and THI monitoring are recommended to improve rabbit productivity in humid tropics.

Keywords: physiological responses, productivity, rabbit does, reproduction performance, temperature-humidity index

1. Introduction

Rabbit production is an important component of smallholder livestock farming systems in Nigeria because of its low capital requirements, fast reproductive cycle, and ability to provide high-quality animal protein for rural and peri-urban households [1, 2]. Rabbits also play a significant role in livelihood diversification by generating supplementary income for farmers through the production of meat, skin, and manure [3]. Despite this potential, rabbit production in the humid tropics remains vulnerable to environmental stressors, particularly temperature and humidity fluctuations, which affect animal health, productivity, and reproductive efficiency [4, 5].

The temperature-humidity index (THI) is widely recognized as a reliable indicator of heat stress in livestock [6]. It integrates ambient temperature and relative humidity to quantify the degree of thermal load experienced by animals [7]. Elevated THI levels disrupt thermoregulation, leading to physiological changes such as increased rectal temperature (RT), respiratory rate (RR), and pulse rate (PR), which negatively influence feed intake, metabolism, and growth performance [5, 8]. These physiological alterations often translate into poor reproductive outcomes, including reduced conception rates (CR), extended gestation lengths, smaller litter sizes, and higher pre- and post-weaning mortality rates [4, 6].

Benin City, located in the humid tropical zone of southern Nigeria, experiences marked seasonal variation between the dry (November–March) and rainy (April–October) seasons [2]. These climatic fluctuations influence THI and, consequently, animal performance. Studies on other small livestock species in Nigeria have highlighted the adverse effects of high THI on productivity [9, 10]. However, there is a paucity of data on how these seasonal changes specifically affect rabbit does, particularly in Benin City, where climatic conditions are characterized by high ambient temperatures and humidity.

Understanding the seasonal variations in THI and their effects on the physiological, reproductive, and productivity performance of rabbit does is essential for improving management practices, designing climate-smart housing, and implementing breeding strategies adapted to local conditions [11, 12]. This study was, therefore, conducted to evaluate how seasonal variations in THI influence the physiological parameters, reproductive traits, and productivity performance of rabbit does raised in Benin City, Nigeria. The findings will provide a basis for the adoption of innovative digital tools and adaptive technologies to reduce heat stress and improve rabbit production in humid tropical environments.

2. Materials and methods

2.1 Experimental site

The experiment was conducted at the Teaching and Research Farm of the University of Benin, Edo State, Nigeria. The University of Benin is located at latitude 6.02°N and longitude 5.06°E in the humid rain forest zone of Southern Nigeria, with an annual temperature range between 24.5°C and 32.7°C, and a mean of 28.6°C. Annual rainfall ranges from 1,498 mm to 3,574 mm with a mean of 2,430 mm. The relative humidity and daily sunshine are between 63.3% and 81.71% and 5.85 hours and 7.5 hours, with means of 73.5% and 6.68 hours respectively (Meteorological Section of the Nigeria Airports Authority [13], Benin City, Edo State, Nigeria).

2.2 Experimental animals

A total of 60 composite rabbits (50 does and 10 bucks) selected from the rabbit population on the farm were used for the study [14]. "Composite rabbit" simply means a breed or strain that resulted from the random crossing of two or more breeds or strains over time, which are found to be geographically adapted, or as described by [15], a non-standard breed of rabbits. However, only the composite rabbit does were monitored in this study.

2.3 Housing, equipment, and management practices

Bucks and does were individually housed in single-tier wooden hutches (60 × 90 × 60 cm) raised 90 cm above the ground to ensure proper sanitation and ventilation. The floors and sides were fitted with wire mesh to allow free droppings, adequate air circulation, and easy observation. Each hutch was equipped with an earthenware feeder and drinker, while nest boxes (30 × 45 × 30 cm) were provided 25 days post-mating. The hutches were positioned to shield the rabbits from rain, wind, and direct sunlight, and all entry points were secured against predators.

Routine hygiene included the twice-daily provision of clean water, daily pen cleaning, and minimal visits outside for feeding and cleaning. Sick animals were promptly treated, and mortalities were removed and buried. A first aid box, buckets, and sanitation tools were kept on-site. Feed and water were supplied ad libitum: commercial grower mash in the morning and forages (*Panicum maximum* and/or *Vernonia amygdalina*) in the afternoon. The diet met NRC (1977) nutrient standards [16–18].

Mating was conducted by natural service, with the doe introduced to the buck's hutch either in the morning (07:00–08:00) or in the evening (16:00). Bucks were used no more than 3–4 days per week and up to two times per day. A semi-intensive reproductive system (10–14 days post-kindling to the next mating) was adopted [19]. Pregnancy was diagnosed through abdominal palpation 10–14 days post-coitus. Nest boxes were checked after kindling for stillbirths and fetal sacs. Weaning was carried out at four weeks, with weaned does transferred to separate hutches. Cage identification was maintained throughout the study.

2.4 Data collection

Data on temperature, relative humidity, physiological parameters, and reproductive traits of the rabbits were collected over a period of one year (January 2013–December 2013), while growth parameters of kits were collected from February 2013 to March 2014 to determine the effect of heat stress on the performance of rabbits.

2.4.1 Daily environmental data

2.4.1.1 Temperature

Temperature was measured twice daily at 09:00 and 14:00 hours using a mercury-in-glass thermometer at six fixed microclimate points (entrance, far corner, central aisle, roof level, top-tier cage front, bottom-tier cage front), all at animal head height except for the roof point. For each time point, the temperature readings from all six locations were summed and divided by six to obtain the mean temperature at that time. The daily average temperature for the rabbitry was then calculated by taking the mean of the two time-point averages each day.

2.4.1.2 Relative humidity

Relative humidity was measured twice daily at 09:00 and 14:00 hours using a wet–dry bulb thermometer (distilled water wick; shaded, aspirated) at the same six fixed microclimate points. The relative humidity readings from these points at each time were summed and divided by six to obtain the mean for that time. The daily average relative humidity was then calculated as the mean of the two time-point means. A wet–dry bulb thermometer and a humidity table conversion chart were used to convert degrees Celsius (°C) readings to percentage (%) values for relative humidity.

To avoid radiant bias, thermometers were positioned more than 30 cm away from walls and metal surfaces, and measurements were consistently taken at the same fixed points each day.

Temperature reading uncertainty was ± 0.5 °C (post-check); RH combined uncertainty was estimated at ± 3% RH by error propagation of dry/wet-bulb readings, respectively.

2.4.2 Physiological data

2.4.2.1 Rectal temperature (RT)

This is the degree of hotness or coldness of the rectum of the experimental animals. This measurement was taken using the Kris Alloy digital thermometer and a stopwatch. It was performed by inserting a digital thermometer into the rectum of the rabbits for one minute, after which the reading was recorded.

2.4.2.2 Respiratory rate (RR)

This was determined by visually counting nasal movements per one minute using a stopwatch.

2.4.2.3 Pulse rate (PR)

This was determined for each animal by placing the left arm between the radial and carotid arteries and counting the contractions of the ventricle and the corresponding number of times the heart beats per one minute.

These readings were taken twice, and the average was recorded.

2.4.3 Reproduction and growth data included

Weight of doe before service or mating (WBSV); date of service (DTSV); number of times served (NTSV); kindling date (KINDT); gestation length (GLT); litter size at birth (LTSB); litter size at weaning (LTSW); records of pre-weaning mortality (PreMor); parity (PRTY); Weekly body weight from 2 weeks to 22 weeks of age, measured using a weighing scale with a 10 kg capacity. Hence, the estimated average body weight in kilograms (kg) would be reported in grams (g), equivalent to the least value in this study.

2.5 Determination of heat stress index (HSI)

However, monthly and seasonal heat stress index (HSI) values were determined using the same procedure as described in [20], which was modified by for rabbits, based on the thermal comfort level of an animal's environment. According to [21], the HSI is a function of the temperature-relative humidity index (THI). The THI was measured according to the following equation [4]:

where t°c = dry bulb temperature in degrees Celsius, and RH = relative humidity as a percentage divided by 100.

$$THI = t - \left[\left(0.31 - 0.31\left(\frac{RH}{100}\right)\right)(t - 14.4)\right] \quad (1)$$

THI values were classified as follows:

<27.8 = absence of heat stress;

27.8–28.9 = moderate heat stress;

29.0–30.0 = severe heat stress; and

> 30.0 = very severe heat stress.

2.6 Statistical analysis

The data obtained were subjected to statistical analysis using the General Linear Model (GLM) procedure of [22] to obtain descriptive statistics. Inferential statistics were conducted by subjecting the data to ANOVA to determine the influence of the THI on the performance of rabbits. Significant means were separated using the Student–Newman–Keuls test.

3. Results and discussion

Table 1 shows the influence of seasonal THI on the physiological responses of does. The highest THI occurred during the late dry season (32.0 ± 0.14), and the

Season	Seasonal THI	RT (°C)	PR (bpm)	RR (nmpm)
Late dry	32.0 ± 0.14^{a}	38.37 ± 0.05^{c}	105.44 ± 1.09^{c}	100.56 ± 1.11^{c}
Early rainy	30.9 ± 0.19^{bc}	38.83 ± 0.04^{a}	122.70 ± 0.61^{b}	115.23 ± 0.70^{a}
Late rainy	29.0 ± 0.19^{d}	38.45 ± 0.02^{b}	124.50 ± 0.42^{a}	111.39 ± 0.49^{b}
Early dry	31.0 ± 0.16^{b}	37.83 ± 0.04^{d}	84.36 ± 0.33^{d}	80.22 ± 0.32^{d}

Values are least square means (± SEM), n = 4.
^{abcd} means that within the same column, values with different superscripts are significantly different (P <0.05).
BPM: beats per minute, NMPM: nasal movements per minute.

Table 1.
Mean values of the breeding does physiological parameters as influenced by seasonal THI.

lowest during the late rainy season (29.0 ± 0.19). RT peaked in the early rainy season (38.83 ± 0.04 °C), while the lowest was in the early dry season (37.83 ± 0.04 °C). PR and RR were also elevated in the rainy seasons, with the highest RR recorded in the late rainy season (124.50 ± 0.42 nmpm). The lowest PR (84.36 ± 0.33 bpm) and RR (80.22 ± 0.32 nmpm) were observed in the early dry season.

The elevated RT, PR, and RR during the rainy seasons reflect adaptive physiological responses to maintain thermal balance under fluctuating humidity and temperatures. Similar findings have been reported by [4] and [5], who noted that heat stress above THI thresholds (30–32) compromises rabbit thermoregulation. Conversely, lower values in the early dry season indicate improved thermal comfort and reduced metabolic stress, consistent with [8].

These results underscore the need for climate-smart housing and real-time THI monitoring to reduce heat stress and maintain optimal productivity in breeding does [11,12].

Table 2 presents the seasonal THI, which significantly ($P < 0.05$) affected the reproductive traits of does. The longest gestation length (GLT) was recorded during the late dry season (32.10 ± 0.23 days), while the shortest occurred in the late rainy season (29.60 ± 0.30 days). CR was lowest in the late dry season (20%) and highest during the early dry season (60%). Similarly, the total number of kits at weaning (TNKW) and at sexual maturity (TNKSM) were lowest in the late dry season (66 and 40, respectively) but highest in the late rainy season (170 and 126, respectively). PWMor was markedly higher in the late dry season (73.08%) compared to the early rainy season (34.88%).

These results indicate that high THI during the late dry season (32.00 ± 0.14) adversely affects conception and kit survival. Heat stress is known to reduce ovarian activity, embryo survival, and milk production, thereby increasing mortality [4, 5]. Conversely, the relatively lower THI of the late rainy season (29.00 ± 0.19) supported better reproductive performance, consistent with reports that moderate thermal conditions enhance conception rates and kit growth [6, 8].

The findings underscore the importance of climate-smart housing and breeding schedules targeted to cooler seasons to improve reproductive efficiency in rabbit production. Integration of real-time THI monitoring and digital reproductive data management could help farmers predict and mitigate seasonal reproductive losses [5, 11, 12].

Table 3 reveals seasonal and parity influences on productivity. CR were lowest during the late dry season (20% in 1st parity, 0% in 2nd parity), coinciding with the highest THI (32.0). Productivity improved markedly from the 4th parity in the early rainy season, with CR reaching 68% and shorter kindling intervals (KI) of 49 days, consistent throughout the late rainy and early dry seasons. The number of kits reaching sexual maturity (GK) was also highest in the early rainy season (49) and lowest during periods of high THI (0–1 kits in 2nd–3rd parities).

The reduced conception and kit survival in early parities under high THI likely result from heat-induced hormonal imbalances, poor ovulation, and increased embryonic mortality [4, 6]. Improved productivity from mid to later parities during moderate THI seasons (rainy and early dry) reflects better reproductive recovery and kit viability, similar to trends reported by [5].

These findings underscore the need for breeding schedules that avoid peak-heat periods, as well as the adoption of climate-smart housing and digital THI monitoring to support reproductive efficiency in rabbit production [11, 12].

DOI: http://dx.doi.org/10.5772/intechopen.1012861

Season	Season THI	Number of does	Reproductive variables								
			GLT (days)	TNKB	LTSB	LTSW	TNKW	CR (%)	PWMor (%)	PostWMor (%)	TNKSM (%)
Late dry	32.00 ± 0.14^{a}	50	32.10 ± 0.23	104^{d}	5.20 ± 0.39	4.13 ± 0.40	66^{c}	20(10/50)^{c}	73.08^{a}	78.78^{a}	40^{c}
Early rain	30.90 ± 019^{bc}	50	31.10 ± 0.33	172^{c}	4.10 ± 0.23	3.74 ± 0.18	142^{b}	42(21/50)^{b}	34.88^{c}	28.18^{c}	122^{a}
Late rain	29.00 ± 0.19^{d}	50	29.60 ± 0.30	236^{b}	4.37 ± 0.27	3.40 ± 0.20	170^{a}	54(27/50)^{ab}	55.94^{b}	51.76^{b}	126^{a}
Early dry	31.00 ± 0.16^{b}	50	30.30 ± 0.40	256^{a}	4.27 ± 0.24	3.60 ± 0.30	144^{b}	60(30/50)^{a}	87.5^{a}	77.78^{a}	88^{b}

Values are least square means (± SEM).
^{abc}*Means within the same row having different superscripts are significantly different ($P < 0.05$). GLT: average gestation length, TNKB: total number of kits born, LTSB: litter size at birth, LTSW: litter size at weaning, TNKW: total number of kits at weaning, CR: conception rate, PWMor: pre-weaning mortality, PostWMor: post-weaning mortality, TNKSM: total number of kits at sexual maturity.*

Table 2.
Reproductive traits of does as influenced by season and THI.

Season	Parity	N	SMD	CR (%)	KI (Days)	GK
Late dry	1st	50	10	20	–	20
	2nd	50	4	0	–	0
Early rain	3rd	50	4	16	98	1
	4th	50	17	68	49	49
Late rain	5th	50	13	52	49	29
	6th	50	14	56	49	34
Early dry	7th	50	13	52	49	26
	8th	50	17	68	49	18

N: number of does; SMD: successive mated doe; CR: conception rate; KI: kindling interval; GK: group of kits that reached sexual maturity at 22 weeks

Table 3.
Productivity performance of does within the expected parities as influenced by seasonal THI.

4. Conclusion

This study demonstrated that seasonal fluctuations in the THI significantly influence the physiological responses, reproductive efficiency, and productivity performance of rabbit does in Benin City, Nigeria. THI values frequently exceeded the thermal comfort threshold during the late dry season, resulting in elevated rectal temperature, respiratory rate, and pulse rate, which reflected severe heat stress. These adverse conditions led to extended gestation lengths, reduced conception rates, high PreMor, and overall poor kit survival, particularly among early-parity does. Conversely, the relatively moderate THI values observed during the late rainy and early dry seasons supported better reproductive outcomes, higher kit survival, and improved productivity.

The findings underscore the vulnerability of rabbit production in humid tropical climates to heat stress and highlight the necessity of climate-adaptive management strategies.

Key recommendations include the adoption of climate-smart housing designs that enhance ventilation and thermal comfort, strategic breeding schedules aligned with cooler seasons, and the use of real-time THI monitoring tools to predict and mitigate the effects of heat stress. Implementing these interventions will improve reproductive efficiency, reduce kit losses, and enhance overall productivity and profitability for smallholder rabbit producers in the humid tropics.

Author details

O.D. Asemota[1,3]*, O.E. Okhale[2,3] and A.M. Orheruata[3]

1 Department of Farming Systems and Research Programme Outreach, National Root Crops Research Institute, Umudike, Nigeria

2 Department of Animal Production, Faculty of Agriculture, Southern Delta University, Ozoro, Nigeria

3 Department of Animal Science, Faculty of Agriculture, University of Benin, Benin City, Nigeria

*Address all correspondence to: aseadonis@yahoo.com

References

[1] Adeniji CA. Adoption of organic poultry and Rabbit production for improvement of quality of life in Nigeria: A review. The International Journal of Organic Agriculture Research and Development. 2019;**15**:35–41. Available from: https://ijoardjournal.org/index.php/ijoardjournal/article/view/150

[2] Ajayi BA, Oseni SO, Adeniyi BD, Ajala, AO, Olubamigbe, JO. Characterisation and needs assessment of rabbit production in Southern Kwarw, Nigeria. Tropical Animal Production Investigations. 2019;**22**(2):24–29. Available from: https://tapianimalsci.ui.edu.ng/index.php/journal/artical/view/174

[3] Ayeni MD, Adewumi MO, Bello MA, AdiAdi KF, Osungade AA. Effects of rabbit production on income and livelihood of rural households in Nigeria. Heliyon. 2023;**9**(8):e18568. DOI: 10.1016/j.heliyon.2023.e18568

[4] Marai IFM, Ayyat MS, Abd El-Monem UM. Growth performance and reproductive traits at first parity of New Zealand White female rabbits as affected by heat stress and its alleviation, under Egyptian conditions. Tropical Animal Health and Production. 2002;**34**:193–204

[5] Renaudeau D, Collin A, Yahav S, de Basilio V, Gourdine JL, Collier RJ. Adaptation to hot climate and strategies to alleviate heat stress in livestock production. Animal. 2012;**6**(5):707–728

[6] Habeeb AA, Gad AE, Atta MAM. Temperature-humidity indices as indicators to heat stress in livestock. Egyptian Journal of Animal Production. 2018;**55**(2):109–118

[7] Collier RJ, Gebremedhin KG. Thermal biology of domestic animals. Annual Review of Animal Biosciences. 2015;**3**:513–532. DOI: 10.1146/annurev-animal-022114-110659

[8] Okab AB, El-Banna SG, Koriem AA. Influence of environmental temperatures on some physiological and biochemical parameters of New-Zealand rabbit males. Slovak Journal of Animal Science. 2008;**41**(1):12–19

[9] Ayo JO, Obidi JA, Rekwot PI. Effects of heat stress on the wellbeing, fertility and hatchability of chickens in the Northern Guinea Savannah zone of Nigeria: A review. Journal of Poultry Science and Technology. 2014;**2**(4):56–62

[10] Ewuola EO, Sanni RO, Aina T, Oni O, Jimoh OA. Theromregulatory response and oxidative Stress indices of rabbit bucks administered ascorbic acid and sodium bicarbonate in a humid tropical environment. The Journal of the Basic and Applied Zoology; 83(55):1–11. DOI: 10.1186/s41936-022-00307-5

[11] Mutwedu VB, Nyongesa AW, Kitaa JM, Ayagirwe RBB, Baharanyi C, Mbaria JM. Effects of moringa aleifera aqueous seed extracts on reproductive traits of heat-stressed New Zealand white female rabbits. Frontiers in Veterinary Science. 2022;**9**:883976. Available from: https://pubmed.ncbi.nlm.nih.gov/36172616?

[12] FAO. The State of Food and Agriculture 2021: Making Agrifood Systems More Resilient to Shocks and Stresses. Rome: Food and Agriculture Organization of the United Nations; 2021

[13] Asemota OD, Aduba P, Bello-Onaghise G, Orheruata A.M. Effect of Temperature-Humidity Index (THI) on the performance of rabbits (*Oryctolagus cunniculus)* in the humid tropics. *Archivos de Zootecnia* 66, no. 254 (2017):257–261. DOI: 10.5772/intechopen.92622

[14] Asemota OD, Aduba P, Bello-Onaghise G, Orheruata AM. Effect of Temperature-Humidity Index (THI) on the performance of rabbits (*Oryctolagus cunniculus)* in the humid tropics. Archivos de Zootecnia. 2017;**66** (254):257–261

[15] Lukefahr SD. Rabbit Project Development Strategies. 1st ed. Texas: Southern University; 1998

[16] Fraga MJ Rabbit production in hot climates. In: Proceedings of the 6th World Rabbit Congress; 1998; Toulouse, France. p. 13–25

[17] Ladokun AO, Egbunike GN. Effect of dietary Crude Protein on the growth rate and digestibility in the rabbit. Proceedings 3rd Annual Conference of Animal Science Association of Nigeria, Lagos 1998 pp. 200–202. Available from: https://funaab.edu.ng/ladokun-a-o-conferences-attended/?utm

[18] Bamikole MA, Ikhatua MI, Ikhatua UJ, Ezenwa IV. Nutritive value of Mulberry (*Morus spp.*) leaves in growing rabbits in Nigeria. Pakisan Journal of Nutrition, 2005;**4**:231–236 DOI: 10.3923/pjn.2005.231.236

[19] Lebas F, Coudert P, de Rochambeau H, Thébault RG. The Rabbit – Husbandry, Health and Production. In: FAO Animal Production and Health Series. Rome, Italy: Food and Agriculture Organization of the United Nations (FAO); 1997. p. 241

[20] Ogunjimi LAO, Ogunwande GA, Osunade JA. "Influence of Building Environment on Rabbit Weight Gain, Feed Efficiency, Rectal Temperature and Respiration Rate in the Humid Tropical Climate of Southwestern Nigeria.". Agricultural Engineering International: the CIGR Ejournal. Manuscript BC 07 008. Vol. X. January, 2008:1–14. Available from: https://scholar.google.com/citations?view_op=view_citation&hl=en&user=e5syFisAAAAJ&citation_for_view=e5syFisAAAAJ:ufrVoPGSRksC

[21] Asemota DO, Orheruata MA. Effect of the Microclimatic Temperature-Humidity Index (THI) on the Productivity Performance of Rabbit. Intechopen. 2021. DOI: 10.5772/intechopen.92622

[22] GENSAT. General Statistics Analysis Tool – User Manual. Version 3.1. USA: GENSAT Statistical Systems; 2008. Available from: https://vsni.co.uk/software/genstat/?utm

Chapter 5

Contactless Temperature Sensors for Enhancing Thermal Comfort Measurements in Buildings

Fatih Evren

Abstract

This chapter provides an overview of contactless temperature sensing technologies and their role in improving thermal comfort measurement in buildings. It begins with the basics of thermal comfort and the limitations of traditional methods. Then, it introduces infrared radiation and optical sensing principles. This is followed by a detailed discussion of single-channel sensors, thermal cameras, infrared arrays, and the challenges associated with their calibration. The chapter examines the efficacy of these sensors in building environments, highlighting their ability to capture real-time temperature data without physical contact, and evaluates the potential benefits of contactless sensors in optimizing occupant comfort and energy efficiency, thereby contributing to the advancement of smart building technologies. Applications are presented across indoor, outdoor, and occupant-centered contexts, showcasing techniques for mean radiant temperature estimation, air temperature field reconstruction, and physiological monitoring. The chapter emphasizes both the opportunities and constraints of these emerging methods, including accuracy, scalability, integration into smart building systems, and alignment with standards such as ASHRAE 55 and ISO 7726.

Keywords: contactless temperature sensing, mean radiant temperature (MRT), infrared thermography, smart building technologies, thermal comfort measurement

1. Introduction

Thermal comfort is a cornerstone of building performance, directly impacting occupant well-being and productivity. Assessment of thermal conditions in a space for thermal comfort relies on a combination of several environmental factors, namely air temperature, humidity, air velocity, and mean radiant temperature [1, 2]. Among these, accurate temperature assessment – particularly mean radiant temperature (MRT) – is essential but has historically been difficult to measure in a practical, non-intrusive way.

With the rise of smart building technologies, there is an increasing demand for sensing systems that are not only accurate but also cost-effective, scalable, and minimally disruptive to occupants. Contactless temperature sensors, leveraging infrared thermography and other optical technologies, have emerged as a compelling solution [3–5]. This chapter explores their application in enhancing thermal comfort measurement, specifically focusing on mean radiant temperature, which is a key parameter, and highlights their potential to advance energy efficiency and occupant-centric design.

2. Background: thermal comfort and measurement challenges

2.1 Fundamentals of thermal comfort

Thermal comfort is commonly defined as "*that condition of mind which expresses satisfaction with the thermal environment*" [1, 6]. Unlike physical parameters such as temperature or humidity, comfort is a subjective perception – dependent not only on measurable environmental conditions but also on personal factors unique to each occupant. For this reason, the measurement of thermal comfort in a space lies at the intersection of engineering, physiology, and psychology. This chapter solely focuses on the environmental aspect of thermal comfort and, more specifically, explores the optical methods for mean radiant temperature, which is a metric to quantify the thermal interaction between the human body and its surrounding surfaces, and which can be measured over several physical quantities as explained below.

2.1.1 Environmental factors

Four primary environmental variables govern the heat exchange between the human body and its surroundings: air temperature, mean radiant temperature, air speed, and humidity. Air temperature is the dry-bulb temperature of the ambient air surrounding the body. Mean radiant temperature (MRT) is defined as the uniform temperature of an imaginary enclosure in which the radiant heat exchange equals that of the actual environment. Air speed is the movement of air at the occupant level, which can enhance convective heat loss or create drafts. Lastly, humidity is the amount of moisture in the air, influencing evaporative heat loss from the human body [1].

Together, these parameters define the thermal environment and are subject to complex interactions. For example, elevated air speed can compensate for higher air temperatures, while radiant asymmetry may cause discomfort even in otherwise neutral conditions[1].

2.1.2 Personal factors

In addition to environmental influences, two personal variables critically affect thermal sensation: metabolic rate and clothing. Metabolic rate is the rate of energy transformation by the body, from chemical energy to heat or mechanical work, varying with activity level. Resting individuals typically exhibit a metabolic rate of about 1.0 met (58.2 W/m^2) [1, 6, 7]. Activities such as walking, cooking, or heavy labor substantially increase this value.

Clothing insulation, clo, is a unit describing the thermal resistance provided by clothing ensembles [1]. A business suit, for example, corresponds to roughly 1.0

clo, while lighter summer clothing may provide only 0.5 clo [6]. Since clothing and activity are dynamic and culturally influenced, they may introduce variability in comfort assessment across settings and populations.

ASHRAE Standard 55 emphasizes that thermal comfort results from the interaction of individual factors and should never be evaluated in isolation. Indices, such as the Predicted Mean Vote (PMV) and the Predicted Percentage Dissatisfied (PPD), combine these variables to provide quantitative estimates of comfort across populations. Meanwhile, adaptive models account for the influence of outdoor conditions and occupant control in naturally ventilated spaces, recognizing that comfort expectations are shaped by context and acclimatization [1].

2.2 Measurement approaches

2.2.1 Traditional approaches

Historically, building researchers and practitioners have relied on a range of contact-based sensors to measure air temperature in buildings. Thermocouples, resistance temperature detectors (RTDs), thermistors [2, 8], and digital sensors (e.g., DS18B20) are commonly used for air temperature monitoring. For radiant conditions, the most widely applied device is the black-globe thermometer, which estimates mean radiant temperature by measuring the equilibrium temperature of a thermally absorptive hollow sphere containing a temperature sensor at the center. Mean radiant temperature is also measured by net radiometers, which consist of thermopiles and covers that are spectrally selective to measure net incoming thermal radiation for both short and long-wave thermal radiation[2]. While effective in controlled laboratory studies, these traditional instruments face several challenges in real-world building environments:

- Sensors are often intrusive; they need to be installed in visible or inconvenient locations, sometimes interfering with normal building operations or occupant activities.

- Globe thermometers may require many minutes to reach equilibrium; published tests report the typical time to reach equilibrium is on the order of ~10–30 + min for 150 mm globes, depending on air movement and construction [4, 9].

- These sensors can measure only the point of installation, while variability inside a space is also greatly important. Practical deployments typically rely on a limited number of sensors, which may fail to capture the spatial variability of indoor thermal environments.

- Net radiometers are typically expensive and require laboratory-grade data acquisition systems.

These limitations have contributed to a growing recognition that traditional measurement methods, though well-established, are not fully aligned with the dynamic and occupant-centric approaches needed in modern buildings.

2.2.2 Emergence of smart and contactless sensing

In parallel with the digital transformation of buildings, a new generation of smart sensing technologies has emerged. Wireless sensor networks, low-cost microelectronics, and cloud-based data platforms now enable unprecedented levels of data collection and analysis. Within this broader landscape, contactless temperature sensors have gained attention as a promising alternative or complement to traditional approaches for mean radiant temperature measurement. Contactless sensors, particularly those based on infrared thermography and optical methods, offer several advantages:

- They can measure surface or radiant temperatures without requiring physical contact.
- Their response time is often much faster than that of contact-based instruments.
- They can provide spatially rich data, especially when using thermal cameras or sensor arrays.

Despite these benefits, several limitations must be acknowledged. Contactless infrared sensors are sensitive to surface emissivity and may produce biased results in the presence of reflective materials. Their line of sight can be obstructed by furniture or occupants, and thermal cameras require careful calibration to ensure long-term stability. Nevertheless, the emergence of contactless sensing aligns well with the growing demand for occupant-centered building operation, where real-time, fine-grained comfort data can inform adaptive control strategies.

2.3 Mean radiant temperature

Mean radiant temperature (MRT) represents the uniform black-enclosure temperature that would exchange the same net long-wave radiation with a person as the actual surroundings[6]. In practice, MRT can be obtained by: i) combining surface temperatures with view factors, ii) aggregating directionally resolved plane radiant temperatures from net-radiometer measurements, or iii) applying convective corrections to a black-globe thermometer reading [1, 2].

2.3.1 Black globe method

As mentioned in the previous section, the most common method for measuring mean radiant temperature is the use of a black globe sensor. A black globe sensor is a hollow, generally metallic globe that consists of a temperature sensor at the center, as shown in **Figure 1**. Under natural convection, mean radiant temperature can be inferred from the equilibrium globe temperature, $T_{gl,}$ using Eq. (1), where T_{air} is the dry-bulb air temperature in K, D is the globe diameter in m, and ε is the surface emissivity of the black globe sensor. Surface emissivity is typically 0.95 for a black matte surface, and the standard diameter of the sensor is 0.15 m [2].

$$\bar{t}_r = \left[T_{gl}^4 + \frac{0.25 \times 10^8}{\epsilon D}\left(T_{gl} - T_{air}\right)\right]^{\frac{1}{4}} - 273.15 \tag{1}$$

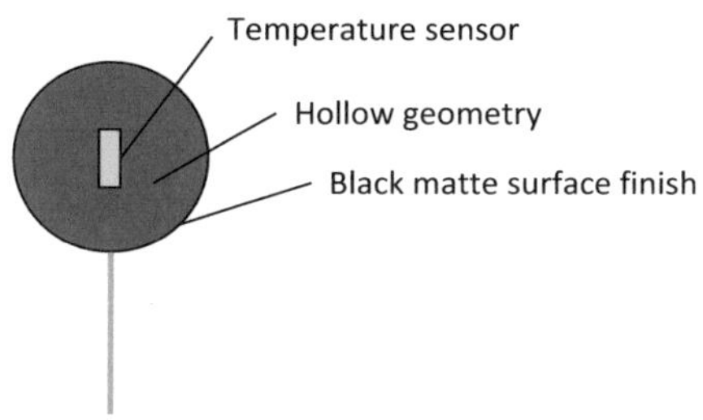

Figure 1.
Black globe sensor schematic. The black globe consists of a hollow sphere with black matte surface finish for high emissivity and a temperature sensor at its center.

When air movement (draught) is present and above a threshold, the influence of forced convection is incorporated through air velocity, v_a in ms^{-1}, and the mean radiant temperature can be calculated using Equation (2) [2].

$$\bar{t}_r = \left[T_{gl}^4 + \frac{1.1 \times 10^8 v_a^{0.6}}{\epsilon D^{0.4}}\left(T_{gl} - T_{air}\right)\right]^{\frac{1}{4}} - 273.15 \tag{2}$$

2.3.2 *Surface temperature method*

When the temperatures of surrounding surfaces, T_i, and the associated view factors, F_{P-i}, between a person and those surfaces are known, MRT can be computed using Eq. (3), where N is the number of surfaces and the temperature is in K. The equation is derived under the assumptions that surfaces behave as gray bodies (their emittance equals their absorptance), that both emission and reflection are diffuse, and that emissivity values are high. Because most building materials meet these criteria, this expression is considered reliable and has become a standard approach for calculation [2, 10]

$$\bar{t}_r = \left(\sum_{i=1}^{N} T_i^4 F_{P-i}\right)^{1/4} - 273.15 \tag{3}$$

For rectangular surfaces, these view factors can either be obtained graphically from reference charts (ISO 7726, Annex B) or computed analytically, as given in the following equations [2, 11].

Eqs. (4–6) [2, 11] define how view factors are determined, with the corresponding fitting coefficients $A, B, C, D, E,$ *and* F_{max} listed in **Table 1**. To apply these relations, the room envelope is subdivided into surface elements relative to the occupant's position and waist height – commonly taken as 1.0 m for a person standing and 0.6 m for a person seated. The geometric parameters a, b, and c are then assigned based on this subdivision. These parameters are shown in **Figure 2**. For horizontal planes such as floors or ceilings, c denotes the vertical distance between the waist level and the surface (e.g., c = 1 m for a person standing on the floor), while a and b describe the length and width of the chosen surface patch aligned with the body's midline. For vertical partitions like walls, a and b likewise represent the

Condition	F_{max}	A	B	C	D	E
Seated (vertical surfaces)	0.118	1.216	0.169	0.717	0.087	0.052
Seated (horizontal surfaces)	0.116	1.396	0.130	0.951	0.080	0.055
Standing (vertical surfaces)	0.120	1.242	0.167	0.616	0.082	0.051
Standing (horizontal surfaces)	0.116	1.595	0.128	1.226	0.046	0.044

Table 1.
Fitting coefficients for view factor calculation equations (2)–(4) [2].

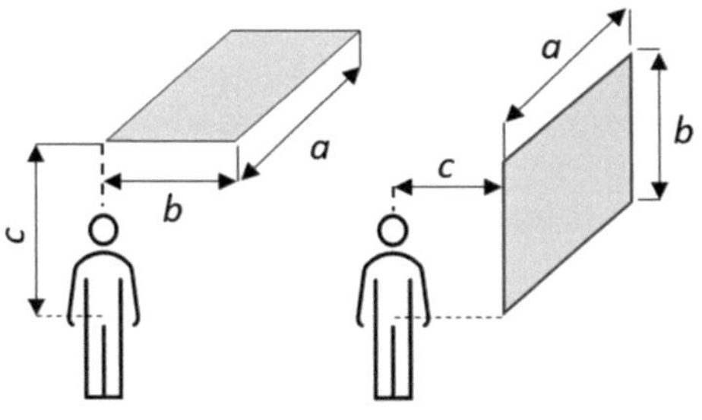

Figure 2.
Distance notation relevant to the human body for MRT calculations for Eqs. (2–4).

dimensions of the subdivided panel at waist height, and c is taken as the horizontal separation between the individual's centerline and the wall surface.

$$F_{p-i} = F_{max}\left(1 - e^{-\frac{a/c}{\tau}}\right)\left(1 - e^{-\frac{b/c}{\gamma}}\right) \tag{4}$$

$$\tau = A + B\left(\frac{a}{c}\right) \tag{5}$$

$$\gamma = C + D\left(\frac{b}{c}\right) + E\left(\frac{a}{c}\right) \tag{6}$$

2.3.3 Plane radiant temperature method

Per ISO 7726 [2], mean radiant temperature can be obtained by combining plane radiant temperatures measured with directional radiometers and weighting them by projected-area factors that represent human body geometry. A net radiometer determines the local radiative balance by sensing both long-wave and short-wave fluxes (typically a pyrgeometer for long-wave and a pyranometer for short-wave) and reporting the net irradiance at the measurement point. Because these measurements are directional, the sensor (or sensor pair) must be aimed along the six orthogonal axes: up, down, left, right, front, and back. For a standing person, when the facing direction is unspecified, the recommended projected-area weights are 0.06 and 0.22, respectively, for horizontal and vertical surfaces. The set of directional plane radiant temperatures is then combined with these weights using the standard MRT weighting formula given in Eq. (3) and can be written as given in Eq. (7) in the explicit form, where $T_{pr\text{-}1}, \ldots, T_{pr\text{-}N}$ are measured plane radiant temperatures for each direction in K, and $F_{p\text{-}1}, \ldots, F_{p\text{-}N}$ are the projected-area weights mentioned above.

$$\bar{t}_r = T_{pr-1}^4 F_{p-1} + T_{pr-2}^4 F_{p-2} + \ldots + T_{pr-N}^4 F_{p-N} - 273.15 \tag{7}$$

In Eq. (7), the plane radiant temperature for a given direction can be calculated using Eq. (8), where $T_{pr\text{-}i}$ is the plane radiant temperature in the ith direction in K, P_i denotes the radiant flux incident from that direction, recorded by the radiometer in units of Wm^{-2}, T_n is the temperature of the radiometer's sensing element in K for that particular direction, and σ is the Stefan–Boltzmann constant (5.6704×10^{-8} $Wm^{-2}K^{-4}$). It should be noted that the precise mathematical form may differ depending on how the radiometer outputs its measurements, since some radiometers give the net radiation, while others give the incoming radiation. Additional details can be found in ISO 7726 for different radiometer configurations [2].

$$T_{pr-i} = \left(0.95\, T_n^4 + \frac{P_i}{\sigma}\right)^{\frac{1}{4}} \tag{8}$$

3. Principles of contactless temperature sensing

3.1 Fundamentals of infrared radiation

Every object with a temperature above absolute zero emits electromagnetic radiation [12]. Infrared (IR) radiation is defined as the spectral region of the electromagnetic spectrum adjacent to visible light, with wavelengths from about 0.78 μm to 1 mm [13].

A blackbody is an ideal emitter/absorber whose spectral radiance depends only on temperature. Real surfaces deviate from blackbodies. The emissivity, ε, describes how closely a surface emits relative to a blackbody and depends on wavelength, temperature, surface finish, and viewing angle. A gray surface, on the other hand, emits and absorbs radiation at a constant fraction of a blackbody, regardless of wavelength. For a gray surface, the thermal radiation emitted is calculated using the Stefan-Boltzmann Law as given in Eq. (9), where E is the total emitted thermal radiation, σ is the Stefan–Boltzmann constant, and T is the absolute temperature in Kelvin [12]. Most building materials exhibit radiative properties that approximate those of a gray surface and diffuse [10, 14].

$$E = \epsilon \sigma T^4 \tag{9}$$

The key concept underlying contactless sensing is that surface temperature can be inferred by measuring the emitted infrared radiation. However, two factors complicate this translation: emissivity and reflected radiation from the surfaces. These considerations mean that contactless sensors must be carefully calibrated and interpreted to achieve accurate results.

3.2 Geometrical transformations

Optical methods for thermal comfort assessment often require mapping thermal images to real-world geometry so that surface temperatures can be associated with their correct spatial location. This is achieved through projective transformations, also known as homography as shown in **Figure 3**, which align points in an image plane with their corresponding points in the physical environment. In homogeneous coordinates, the relationship between a point before transformation $\tilde{x}$, and its transformed counterpart $\tilde{x}$ is expressed as given in Eq. (10), where $\tilde{H}$ is a 3 × 3 homography matrix and tilde, ~, shows the homogenous coordinates [15].

$$\tilde{x}' = \tilde{H}\tilde{x} \tag{10}$$

Because a projective transformation has eight degrees of freedom, a minimum of four pairs of corresponding points between the thermal image and the reference geometry are required to solve for the matrix. Once estimated, the transformed points must be normalized to obtain the inhomogeneous coordinates using Equations [15, 16]:

$$x' = \frac{h_{11}x + h_{12}y + h_{13}}{h_{31}x + h_{32}y + h_{33}} \tag{11}$$

$$y' = \frac{h_{21}x + h_{22}y + h_{23}}{h_{31}x + h_{32}y + h_{33}} \tag{12}$$

Where h_{11}, h_{12}, ... h_{33} are the elements of the homography matrix. These transformations preserve straight lines but not lengths, angles, or parallelism; thus, in the application, these parameters should be considered [17].

Traditionally, stereo calibration is often used in conjunction with projective transformations. Stereo systems employ two cameras – either thermal or thermal paired with RGB – to establish 3D geometry through triangulation. By identifying corresponding feature points across both image planes, the relative orientation and intrinsic parameters of the cameras can be calibrated [15]. Recent advancements also allow a single camera, low-resolution thermal array sensor to obtain the calibration data [4].

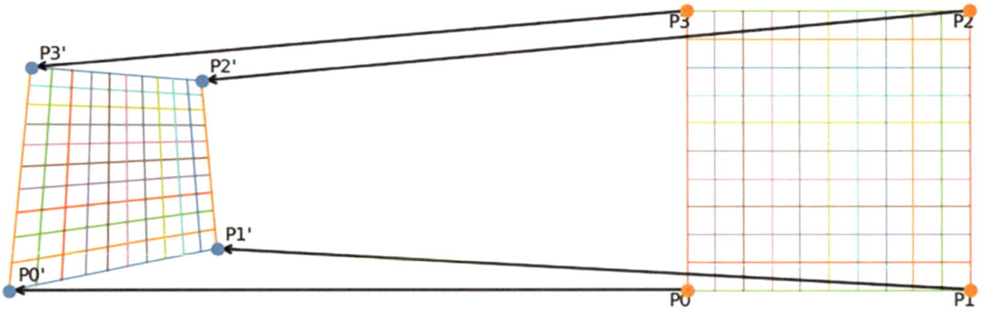

Figure 3.
Schematic representation of geometric transformation. Original points (P0, P1, P2, P3) are transformed to the transformed points (P0', P1', P2', P3'), using the homography matrix $\boldsymbol{H}$

3.3 Types of contactless temperature sensors

A variety of sensor technologies have been developed to measure temperature without physical contact. Sensors used for thermal comfort measurements can be broadly classified into three categories based on their pixel resolution:

Infrared Thermopile Sensors: thermopile sensors measure the infrared radiation from a specific point or narrow field of view. They are relatively inexpensive, compact, and easily integrated. Their low cost makes them suitable for widespread deployment, but they provide limited spatial information.

Infrared Array Sensors: infrared array sensors contain multiple thermopile elements or MEMS sensor arrays arranged in a matrix, providing coarse spatial resolution (e.g., 8 × 8 or 16 × 16 pixels). These sensors can capture thermal distributions across a small area, offering more context than single-point devices while still being affordable and power-efficient.

Thermal Imaging Cameras: at the higher end of complexity and cost are thermal cameras, which use focal-plane array detectors to produce detailed images of thermal radiation. These devices provide rich spatial data for use in research. Their expense, form factors, and processing requirements currently limit routine deployment.

3.4 Calibration and accuracy considerations

To ensure reliability, contactless temperature sensors must address several sources of error. i) Emissivity correction is required. Many sensors allow users to set emissivity values to match the target material, but incorrect assumptions can introduce errors of several degrees Celsius. In the building science domain, it is typically assumed that the emissivity of most common materials is above 0.9. ii) Distance and field of view are critical factors to consider. Infrared sensors have a distance-to-spot size ratio; beyond a certain range, they average over larger areas, reducing precision. In addition, if the FOV is not sufficient to capture critical surfaces, it typically requires a pan/tilt mechanism to measure the mean radiant temperature using optical sensors. iii) Environmental influences are important; for instance, reflective surfaces can transmit or reflect IR radiation, leading to erroneous results. Despite these challenges, recent studies have shown that, when properly deployed, contactless sensors can achieve accuracy within ±0.5–1.0°C compared to reference methods [4].

4. Applications of contactless temperature sensing for indoor environmental quality measurements

4.1 Indoor air temperature distribution measurements

Infrared (IR) thermography can be applied to characterize indoor air temperature distributions by introducing mock targets into the environment. Because air is transparent in the infrared spectrum, auxiliary targets with known thermal properties are required to exchange heat with the surrounding air, thereby enabling indirect measurement. The surface temperature of these targets is monitored with an IR camera, and air temperature and velocity fields are reconstructed by solving corresponding heat transfer models [18–20]. A schematic view of this method is presented in **Figure 4.**

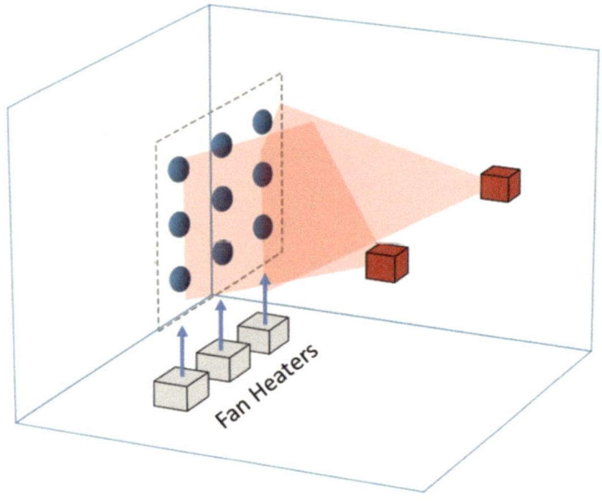

Figure 4.
Schematics of mock target IR thermography. Mock target typically consists of an array of objects, preferably spheres to ensure equal thermal interaction in all directions, to capture spatial differences.

One implementation uses mock target grids composed of small spheres in combination with stereo-calibrated infrared cameras. To separate convective and radiative contributions, the grids are equipped with mini-fans that create two distinct steady states. This setup allows the calculation of two and three-dimensional air temperature and velocity maps. This method was tested in the field [3, 20] and in a controlled test chamber, achieving deviations of less than 1% compared to thermocouple measurements. Furthermore, it showed that PVC mock targets provided faster equilibration and more reliable results than metallic alternatives [19].

Although these methods achieve high spatial resolution and can reduce reliance on dense thermocouple networks, they require stereo calibration, auxiliary equipment, and high computational effort. The necessity of multiple mock target layers and forced airflow introduces significant energy demand. Consequently, while mock target IR thermography is a valuable laboratory tool, its intrusive nature prevents its use in occupied indoor environments.

4.2 Mean radiant temperature measurements

4.2.1 Single channel contactless sensors

Single-channel infrared (IR) sensors, such as thermopiles, pyrometers, and compact IR modules, provide a direct and non-intrusive means of estimating mean radiant temperature (MRT). Their compact size and low power demand make them suitable for both laboratory studies and practical building applications.

The fundamental principle is straightforward: a single IR sensor detects longwave radiation from its field of view and, when combined with surface emissivity assumptions and view factor models, provides an estimate of the surrounding radiant environment. Studies have demonstrated that such sensors can reproduce MRT within ±0.5°C of reference values while substantially reducing response time [21–25].

A practical implementation of a single-channel contactless sensor employs a thermopile mounted on a rotating platform to sequentially scan the surrounding surfaces. During rotation, the device records both the surface temperature and the angular position, and, in some cases, the distance to the surface, creating a directional map of the enclosure. The collected data are processed using view factor algorithms to weight each directional measurement, allowing the computation of mean radiant temperature [21–25].

Despite its advantages, this approach presents several limitations. The reliance on a pan–tilt mechanism introduces moving parts, which increase mechanical complexity and reduce long-term reliability. Because the sensor interprets all radiant sources as part of the enclosure, the presence of occupants or transient heat sources can distort the measurements unless additional filtering is applied. The sequential scanning process also requires several minutes to complete a full cycle, which makes the system less suitable for capturing rapid transients in the thermal environment. Finally, the need for geometric reconstruction and view factor calculations increases computational demand compared to simpler point-based sensors.

Single-channel systems can also be used as an array to capture radiant asymmetry in a space and as an alternative to a globe sensor. Instead of relying on the thermal equilibrium of a blackened sphere, these instruments detect longwave radiation from surrounding surfaces within their field of view. A representative approach uses micro-thermopile sensors embedded on the faces of a geometric solid, such as a cube or icosahedron, to capture directional radiative flux. Each sensor records the radiant exchange from a specific orientation, and the combined measurements are integrated using projected area factors to yield MRT [26]. This configuration remains a point-based measurement, representing conditions only at the sensor's location, and while it provides real-time and directionally resolved data, it cannot capture spatial variability across a room.

4.2.2 Thermal camera applications

Thermal cameras, unlike single-channel sensors, can record spatially resolved surface temperatures across an entire field of view in a single frame. By combining these data with geometric models of the space, MRT can be calculated for one or more occupant locations without requiring intrusive instruments to be installed at the point of interest.

One approach is to use one or more infrared cameras that are manually rotated or repositioned to capture multiple thermal images of an interior space, as shown in **Figure 5**. These images are then corrected for perspective distortion and stitched together into a continuous thermal map. The stitched data can be aligned with simplified geometry or BIM models to represent the distribution of surface temperatures across walls, ceilings, and floors. From these composite images, angle factors are applied to weight each surface's contribution to the occupant position, enabling the calculation of mean radiant temperature. This method is practical for covering large or complex spaces that cannot be captured in a single thermal frame, though it requires careful calibration, consistent emissivity settings, and post-processing effort to ensure accuracy [27, 28].

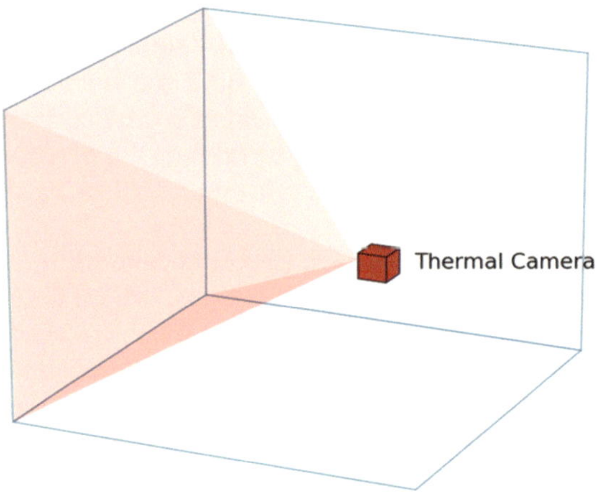

Figure 5.
Schematic representation of the thermal camera method with the field of view where the thermal camera measures the thermal distribution of the surface within its field of view, as shown in red. It needs to be rotated in different directions to capture the entire surrounding surface temperature distribution.

To automate the scanning, a common approach integrates a thermal camera with a pan–tilt mechanism, similar to the single-channel scanning tools. Each captured thermal frame is tagged with its angular position, and surface boundaries are classified in advance through simple 3D modeling. After correcting for emissivity, reflection, and transmittance, the average temperature of each surface is derived. These values are then weighted by angle factors to compute MRT at multiple occupant positions. Experimental studies demonstrated that this method can reveal significant spatial variability in radiant environments, with differences in MRT of up to 6°C within a single room, depending on the measurement location [5, 27].

Recent work has extended this principle to include both longwave and shortwave components. In such systems, a thermal camera records the longwave emission of walls, floors, and ceilings, while a pyranometer measures diffuse and direct solar irradiance entering through transparent façades. The combined data are processed using projected area factors and solar geometry to estimate the contribution of sunlight to the occupant's radiant exposure. This integration enables MRT estimation that accounts for solar radiation effects [29].

Emerging approaches further enhance this capability by combining thermal cameras with other imaging modalities. For example, coupling infrared imaging with LiDAR and high-dynamic-range (HDR) cameras allows systems not only to measure surface temperatures but also to reconstruct the 3D geometry of the space, detect occupants, and evaluate thermal comfort indices such as Predicted Mean Vote (PMV) [30].

Thermal camera–based MRT measurement offers several advantages: rapid acquisition of large areas, integration of solar gains, and the ability to estimate conditions at multiple occupant locations without physical probes. However, these systems also face challenges, including computationally intensive processing and cost.

4.2.3 Low resolution/stationary sensor techniques

Stationary contactless sensing techniques rely on fixed infrared arrays or compact thermal detectors mounted at permanent locations within a room. Unlike scanning systems, which require pan–tilt mechanisms or manual repositioning, stationary sensors continuously monitor their field of view without moving parts. The approach is based on the principle that low-resolution infrared arrays can capture sufficient directional radiative information when sensor fusion is applied and combined with geometric correction algorithms. By applying homography transformations, the sensor output is mapped to the surrounding room surfaces, enabling the reconstruction of directional surface temperatures.

These measurements are then processed using projected area factors to weight each surface according to its contribution to an occupant's exposure. The result is a real-time estimate of mean radiant temperature with accuracy comparable to laboratory-grade radiometers. Validation studies against a six-directional net radiometer confirmed that the stationary IR array technique maintains errors within ±0.5°C, even under conditions of asymmetric radiation and variable surface emissivity [4]. The absence of moving components significantly reduces energy consumption, hardware complexity, and computation requirements, while still achieving continuous temporal resolution.

Stationary IR sensing is particularly well-suited for integration into smart building systems. Compact modules can be ceiling-mounted or wall-mounted to cover representative portions of the enclosure and connected wirelessly to building management platforms. Because the sensors are inexpensive and easily replicated, multi-sensor networks can be deployed to capture radiant asymmetry across different zones of a building.

4.2.4 Physiological parameter measurements

Thermal comfort can also be assessed through physiological parameters, particularly skin temperature, which reflects the body's thermoregulatory response to the surrounding environment. Thermal imagery provides a non-contact approach for measuring skin temperature, offering an alternative to invasive or uncomfortable sensor attachments. By capturing radiative emissions from exposed body regions such as the face, chest, or arms, infrared cameras can estimate surface temperature distributions that correlate with thermal sensation and comfort indices. Studies have demonstrated that averaging skin temperatures across multiple facial regions produces stronger correlations with subjective comfort votes than relying on a single measurement point [31].

In applied settings, thermal imaging has been used both in controlled chambers and in occupied buildings to evaluate comfort directly from skin temperature. For example, infrared cameras have been deployed alongside skin thermistors to validate comfort models, showing close agreement between thermographic estimates and contact measurements, and explored integrating skin temperature mapping into adaptive building control, where local gradients around the face and torso captured by infrared cameras informed adjustments to HVAC operation [32]. While this approach offers rapid, non-intrusive, and spatially rich data, it also requires careful calibration and complex algorithms. Nonetheless, thermal imagery of skin temperature represents a promising direction for occupant-centered comfort assessment, bridging the gap between physiological response and environmental monitoring.

4.2.5 Outdoor thermal comfort measurement applications

As in indoor environments, mean radiant temperature is key to assessing outdoor thermal comfort, but the complexity of urban settings makes its measurement more challenging. Solar radiation, reflections from building façades and pavements, and longwave emissions from surrounding surfaces all contribute to the radiant load on the human body. To address these factors, optical sensor methods such as panoramic infrared thermography and combined infrared–HDR imaging have been developed. These techniques capture directional surface temperatures and solar contributions, which are then weighted by projected area factors to estimate MRT in accordance with international standards [33, 34].

Compared to traditional black globe thermometers, optical methods provide faster response, wider spatial coverage, and the ability to differentiate between shaded and sunlit conditions. This enables more accurate mapping of thermal comfort across streets, plazas, and other outdoor environments, supporting urban design strategies that mitigate heat stress. While these approaches require careful calibration and substantial data processing, they offer a promising path toward high-resolution, non-intrusive outdoor comfort assessment.

5. Conclusion

This chapter presents novel approaches for thermal comfort measurement through optical sensing technologies. Traditional instruments, such as globe thermometers and net radiometers, though inexpensive and widely used, suffer from major drawbacks, including slow response times, intrusiveness, and the inability to capture spatial variability. These shortcomings have historically constrained the integration of mean radiant temperature (MRT) into practical building control strategies. In contrast, optical methods – ranging from single-channel infrared sensors to advanced thermal cameras – have demonstrated the ability to deliver rapid, accurate, and non-intrusive measurements that align well with the demands of modern smart buildings.

Recent advances highlight particularly promising solutions. Stationary low-resolution infrared sensor networks can replicate many of the advantages of thermal cameras at lower cost and computational demand, enabling scalable deployment across building systems. Nevertheless, whether thermal camera or low-resolution optical sensor calibration is important, the typically high emissivity of indoor materials (>0.9) minimizes error concerns under standard window-to-wall ratios. Additionally, optical methods address privacy challenges associated with high-resolution thermal imaging: single-channel and low-resolution sensors provide useful MRT data without exposing occupant details. Taken together, these developments position optical sensing as a robust and practical pathway for integrating MRT into occupant-centric building operations, enabling more responsive, efficient, and comfortable indoor environments.

Contactless temperature sensors are beneficial in all building types for improving thermal comfort measurements. They are particularly beneficial for occupants who are more sensitive to thermal conditions, such as infants. In addition, MRT measurement is critical in poorly insulated buildings since MRT–air temperature differences can be significant in older buildings with poor insulation or forced convective heating, where interior surface temperatures may be much lower than the air temperature. Spaces heated or cooled with large radiant surfaces, such as radiant floor systems, can especially benefit from MRT-based control using contactless sensors, as these systems are primarily designed to satisfy MRT rather than air temperature. In retrofitting existing buildings, challenges include limited mounting locations, obstructions from furniture or occupants, and variations in surface emissivity. Despite these constraints, compact infrared sensors and sensor networks can be integrated with minimal disruption to provide accurate measurements of MRT.

Author details

Fatih Evren
PNNL, Seattle, WA, USA

*Address all correspondence to: fatih.evren@pnnl.gov

References

[1] ANSI/ASHRAE Standard 55-2020, ANSI/ASHRAE Standard 55 - Thermal Environmental Conditions for Human Occupancy, 2021

[2] ISO 7726. Ergonomics of the thermal environment-Instruments for measuring physical quantities. IS/DISO Standard. 2023;**7726**(E)

[3] Dizeu FBD, Maldague X, Bendada A. Mapping of the indoor conditions by infrared thermography. Journal of Imaging. 2016;**2**(2, Art. no. 2). DOI: 10.3390/jimaging2020010

[4] Evren F, Biswas S, Graves R. Measuring mean radiant temperature for indoor comfort assessment using low-resolution optical sensors. Nature Communications. 2025;**16**(1):1215. DOI: 10.1038/s41467-024-55122-z

[5] Lee D-S, Jo J-H. Pan–tilt IR scanning method for the remote measurement of mean radiant temperatures at multi-location in buildings. Remote Sensing. 2021;**13**(11):2158. DOI: 10.3390/rs13112158

[6] ISO Standard 7730, ISO 7730 – 2005 Ergonomics of the thermal environment — Analytical determination and interpretation of thermal comfort using calculation of the PMV and PPD indices and local thermal comfort criteria. 2005

[7] ASHRAE Handbook Fundamentals. Chapter 9 -thermal comfort. ASHRAE Handbook Fundamentals; 2009. p. 9.6

[8] ASHRAE Handbook Fundamentals, "Chapter 36 - MEASUREMENT AND INSTRUMENTS," 2009

[9] d'Ambrosio Alfano FR, Ficco G, Frattolillo A, Palella BI, Riccio G. Mean radiant temperature measurements through small black globes under forced convection conditions. Atmosphere. 2021;**12**(5):621. DOI: 10.3390/atmos12050621

[10] Fanger PO. Thermal Comfort: Analysis and Applications in Environmental Engineering. Copenhagen: Danish Technical Press; 1970

[11] Cannistraro G, Franzitta G, Giaconia C, Rizzo G.Algorithms for the calculation of the view factors between human body and rectangular surfaces in parallelepiped environments. Energy and Buildings. 1992;**19**(1):51–60

[12] Modest MF, Mazumder S. Radiative Heat Transfer. Fourth ed. London: Academic Press; 2022

[13] Vollmer M, Möllmann K-P Infrared Thermal Imaging: Fundamentals, Research and Applications. Second ed. Weinheim: Wiley-VCH Verlag GmbH & Co. KGaA; 2018. DOI: 10.1002/9783527693306

[14] Hart J. A practical guide to infra-red thermography for building surveys. BRE Publication 1991. p. 46

[15] Szeliski R. Geometric primitives and transformations, Computer Vision Algorithms and Applications. Springer, 2011. p. 50

[16] Hartley R, Zisserman A. Multiple View Geometry in Computer Vision. Cambridge: Cambridge University Press; 2003

[17] Forsyth DA, Ponce J. Computer Vision A Modern Approach. Boston: Pearson; 2013

[18] Dizeu FD, Maldague X, Bendada A. Mapping of the Indoor Conditions by Infrared Thermography. Journal of Imaging. 2016;**2**(2):10. DOI: 10.3390/jimaging2020010

[19] Fokaides PA, Jurelionis A, Gagyte L, Kalogirou SA. Mock target IR thermography for indoor air temperature measurement. Applied Energy. 2016;**164**:676–685. DOI: 10.1016/j.apenergy.2015.12.025

[20] Djupkep FBD, Maldague X, Bendada A, Bison P.Analysis of a new method of measurement and visualization of indoor conditions by infrared thermography. Review of Scientific Instruments. 2013;**84** (8, Art. no. 8). DOI: 10.1063/1.4818919

[21] Revel GM, Arnesano M, Pietroni F, Schmidt M, Kaschtschejewa O, "Evaluation in a controlled environment of a low-cost IR sensor for indoor thermal comfort measurement," in *Proceedings of the 2014 International Conference on Quantitative InfraRed Thermography*, QIRT Council, 2014. DOI: 10.21611/qirt.2014.133

[22] Revel GM, Arnesano M, Pietroni F. Development and validation of a low-cost infrared measurement system for real-time monitoring of indoor thermal comfort. Measurement Science and Technology. 2014;**25**(8, Art. no. 8):085101. DOI: 10.1088/0957-0233/25/8/085101

[23] Revel GM, Arnesano M, Pietroni F, "An innovative low cost IR system for real- time measurement of human thermal comfort"

[24] Revel GM, Sabbatini E, Arnesano M. Development and experimental evaluation of a thermography measurement system for real-time monitoring of comfort and heat rate exchange in the built environment. Measurement Science and Technology. 2012;**23**(3):035005. DOI: 10.1088/0957-0233/23/3/035005

[25] Houchois N, Teitelbaum E, Chen KW, Rucewicz S, Meggers F. The SMART sensor: Fully characterizing radiant heat transfer in the built environment. Journal of Physics: Conference Series. 2019;**1343** (1):012073. DOI: 10.1088/1742-6596/1343/1/012073

[26] Fontana L. Study on a mean radiant temperature measure tool based on an almost spherical array of radiometric sensors. Review of Scientific Instruments. 2012;**83**(11):114906. DOI: 10.1063/1.4764534

[27] Lee DS, Kim EJ, Cho YH, Kang JW, Jo JH. A field study on application of infrared thermography for estimating mean radiant temperatures in large stadiums. Energy and Buildings. 2019;**202**. DOI: 10.1016/j.enbuild.2019.109360

[28] Natephra W, Motamedi A, Yabuki N, Fukuda T. Integrating 4D thermal information with BIM for building envelope thermal performance analysis and thermal comfort evaluation in naturally ventilated environments. Building and Environment. 2017;**124**:194–208. DOI: 10.1016/j.buildenv.2017.08.004

[29] Lee D-S, Jo J-H. Application of IR camera and pyranometer for estimation of longwave and shortwave mean radiant temperatures at multiple locations. Building and Environment.

2022;**207**:108423. DOI: 10.1016/j.buildenv.2021.108423

[30] Woods LD, Macatuno F, Ayala N. Integrating IR, HDR and Lidar Cameras into Building Controls. ASHRAE Transactions. 2024;**130**:2024

[31] Almeida RMSF, Barreira E, Simões ML, Sousa TSF. Infrared Thermography to Evaluate Thermal Comfort under Controlled Ambient Conditions. Applied Sciences. 2022;**12**(23):12105. DOI: 10.3390/app122312105

[32] Metzmacher H, Wo¨lki D, Schmidt C, Frisch J, Van Treeck C, "Real-time assessment of human thermal comfort using image recognition in conjunction with a detailed numerical human model," presented at the 2017 Building Simulation Conference, 2017. DOI: 10.26868/25222708.2017.180

[33] Hatefnia N, Barakati A, Ghobad M, Panah AE, "Reliable methodology to monitor and assess radiant environments," in *Proceedings of 33rd PLEA International Conference*, 2017, pp. 1092–1099

[34] Middel A, Huff M, Krayenhoff ES, Udupa A, Schneider FA. PanoMRT: Panoramic infrared thermography to model human thermal exposure and comfort. Science of the Total Environment. 2023;**859**:160301. DOI: 10.1016/j.scitotenv.2022.160301

Chapter 6

Recent Application of Advanced Temperature Sensors in Special Scenarios of Nuclear Energy Research

Haicai Lyu, Xiaoyang Lun, Xianjun Chen, Pengcheng Yang, Mingqiang Yi and Fenglei Niu

Abstract

In nuclear energy research, the precise measurement of coolant temperature under extreme conditions of high temperature, high pressure, and intense radiation is a critical parameter for evaluating nuclear safety design and ensuring reactor efficiency. This section introduces methods for precise temperature measurement in such demanding environments, covering both the essential hardware and the sophisticated software components of modern temperature sensing systems. Specific coolants discussed include advanced options like liquid metals and supercritical carbon dioxide, each posing unique measurement challenges due to their distinct thermophysical and chemical properties. The advanced measurement techniques employed to address these challenges involve specially designed radiation-hardened armored thermocouples, immune fiber-optic temperature sensors, and robust thermistor-based detection systems. The following analysis compares and evaluates the measurement accuracy, long-term stability, and specific installation requirements of these sensor types under intense radiation conditions, noting the distinct advantages and practical considerations of each type—such as the ruggedness of thermocouples versus the electromagnetic immunity of fiber optics. Finally, typical integrated measurement solutions are provided for key operational parameters; these include strategies for measuring the central temperature of reactor fuel elements and for monitoring the coolant temperature at the core inlet and outlet with high reliability. Collectively, these advanced application cases and technological comparisons offer valuable insights and practical references for temperature measurement in nuclear energy research, directly supporting the ongoing safety design and thermal-hydraulic optimization of current and next-generation nuclear systems.

Keywords: temperature sensors, nuclear energy, harsh irradiation environments, high-temperature and high-pressure environments, engineered sheathed thermocouples, fiber optic sensing, liquid crystal thermography

1. Introduction

Accurate temperature measurement within nuclear reactors is critically important for balancing economic efficiency and operational safety. As a key process parameter, temperature monitoring constitutes an essential aspect of reactor status surveillance. It is imperative to maintain all temperature parameters below established safety thresholds to ensure secure and hazard-free operation. Concurrently, optimizing nuclear fuel temperature under safe operating conditions is necessary to maximize fuel utilization efficiency, thereby enhancing overall reactor performance. Consequently, precise temperature monitoring is fundamental to achieving both economic and safety objectives. Significant measurement deviations could potentially lead to excessive core temperatures, potentially triggering severe accidents including core meltdown and radioactive contamination, ultimately resulting in substantial losses. Compared to other harsh environments, nuclear radiation represents a distinctive environmental factor specific to reactors, beyond extreme temperatures and high pressure. Reactors generate energy through fission reactions, which produce a plethora of radioactive species, including alpha particles, neutrons, and gamma rays. When materials are exposed to radiation, their physical, chemical, electrical, and mechanical properties may undergo alterations due to changes in material composition or structure. These modifications can affect the temperature-response characteristics of temperature-sensing elements based on such materials, ultimately leading to measurement inaccuracies. Therefore, temperature sensors used in reactors – including both sensing elements and measurement leads – must possess not only excellent resistance to high temperatures and pressure but also sufficient radiation resistance. Furthermore, as remote measurement is necessary during reactor operation, detection components with electrical signal outputs are highly preferred. Since the sensor probe directly contacts radioactive media, it becomes highly radioactive itself. Consequently, the probe is required to be compact in size, have a smooth surface, allow easy installation and removal, and ensure reliable sealing. These features help minimize radioactivity accumulation, facilitate decontamination, and simplify maintenance procedures.

Addressing the temperature measurement requirements in nuclear energy research, this section will introduce four types of advanced temperature sensors typically employed in nuclear power studies: sheathed thermocouples, fiber-optic temperature sensors, acoustic thermometry sensors, and thermochromic liquid crystal sensors. Each sensor type will first be described in terms of its fundamental measurement principles, followed by an elaboration of its current research status within the nuclear field. Finally, practical application scenarios for each sensor type will be presented.

2. Thermocouple temperature sensors

2.1 Principles of thermocouple

The fundamental principle of thermocouple temperature sensors is based on the thermoelectric effect, where a closed circuit formed by two different metals generates a thermoelectromotive force when the two ends of the conductors are maintained at different temperatures. This thermoelectric effect arises due to the difference in the electronic properties of the two conductor materials, particularly

DOI: http://dx.doi.org/10.5772/intechopen.1013464

their electron density. During operation, one end of the thermocouple is placed in the target environment at temperature, while the other end is kept at a reference temperature t0. The resulting thermoelectric voltage consists of two components: the contact potential, caused by the difference in electron density between the two conductors, and the thermoelectric potential (also referred to as the Seebeck effect), which depends on the intrinsic properties of the conductors and the temperature difference between the two ends.

2.2 Research status and analysis

Since the discovery of the thermoelectric effect by Seebeck in 1821, over 300 materials have been used to form thermocouples, among which eight types have been standardized, such as Type S, T, N, and K. In 1957, the Dutch company Philips developed a cable-like thermocouple by combining Type K thermocouple wires with mineral insulation material (MgO) and an outer stainless steel protective sheath, processed through swaging or drawing to form an integrated thermocomposite material known as a sheathed thermocouple, as shown in **Figure 1**. It is widely used in aviation, aerospace, energy, chemical, mechanical, and other fields, marking a transformative advancement in the history of thermocouple materials. For temperature measurement of the coolant at the outlet of a pressurized water reactor (PWR) core, nuclear-grade Type K (NiCr-NiAl) sheathed thermocouples are typically employed, accompanied by accessories such as junction boxes and compensation cables. Sheathed thermocouples primarily consist of metal sheath material, insulation material, and thermocouple wires.

When a thermocouple is exposed to neutron irradiation, neutrons may cause transmutation of the material composition of the thermocouple elements or induce atomic displacement, altering the material structure. Both effects can change the thermoelectric properties of the material, leading to variations in the thermoelectromotive force of the thermocouple. This creates a deviation from the calibration results of the thermocouple, resulting in temperature reading drift. To address this issue, there are three main solutions: first, select materials with small neutron cross-sections to manufacture thermocouple elements, or optimize the structure of the thermocouple to reduce the probability of neutron interaction with the material; second, conduct theoretical analysis and quantitative research on transmutation to correct the thermocouple readings; third, provide in-situ calibration for the thermocouple.

The Mo–Nb thermocouple is a product developed under the first solution. Its temperature measurement schematic is shown in **Figure 2**, where Mo and Nb serve as the two component wires of the thermocouple. Both Mo and Nb have very small neutron absorption cross-sections, allowing them to maintain material stability even under prolonged exposure to nuclear radiation environments. In 2020, Palmer *et al* [1]. conducted in-reactor testing of a high-temperature irradiation-resistant thermocouple (HTIR-TC) based on Mo–Nb components in an advanced test reactor

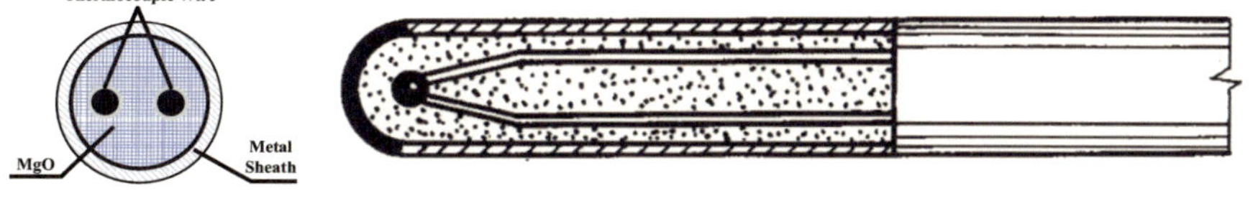

Figure 1.
Structure of a sheathed thermocouple.

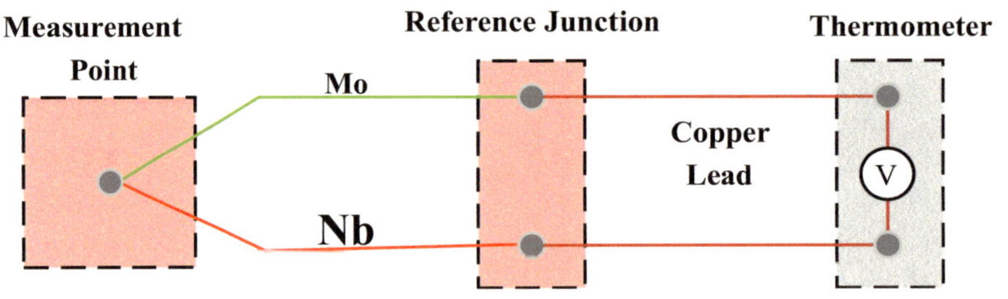

Figure 2.
Schematic diagram of temperature measurement using molybdenum-niobium (Mo–Nb) thermocouple.

(ATR). Over 170 days of irradiation (cumulative thermal neutron fluence < 3.0×10^{21} n/cm^2), the thermocouple exhibited low drift. One of the HTIR-TCs operated stably for 85 days at temperatures between 1,450 °C and 1,500 °C, with deviations from the actual temperature ranging from 30 °C to 50 °C.

For the second approach, nuclear reaction analysis can determine the types and quantities of transmuted elements. Using established correlation curves, the change in thermoelectromotive force caused by elemental variation can be identified and converted into a temperature error, thereby quantitatively estimating the impact of transmutation on thermocouples. Since the 1960s, researchers worldwide have continuously studied the drift behavior of common in-reactor thermocouples under irradiation. In 1962, Ross [2] investigated platinum-rhodium thermocouples by calculating the percentage change in rhodium transmutation based on average neutron energy cross-sections, neutron flux, and exposure time. Combining this with theoretically derived curves, he obtained the theoretical drift in thermoelectromotive force. A comparison between theoretical calculations and experimental results showed high consistency, proving that temperature drift caused by transmutation from nuclear reactions in thermocouples can be predicted through theoretical derivation. In 2013, Scervini *et al* [3]. calculated the composition changes in nickel-based thermocouples, tungsten-based thermocouples (W-5%Re/W-26%Re, W-3%Re/W-25%Re), platinum-based thermocouples (Type S, Pt/Pd), and
Mo–Nb thermocouples after irradiation in PWR and Liquid Metal Fast Breeder Reactor (LMFBR) environments. The results indicated that thermocouples undergo more significant transmutation under thermal neutron irradiation compared to fast neutron irradiation. Nickel-based thermocouples exhibited the smallest compositional changes, while platinum-based and tungsten-based thermocouples experienced considerably greater transmutation. In 2022, Skifton *et al* [4]. developed a drift model for high-temperature irradiation-resistant Mo–Nb thermocouples, accounting for high-temperature, ultra-high-temperature, thermal neutron irradiation, and fast neutron irradiation environments. This model uses correction parameters derived from extensive thermocouple irradiation test data to predict the effects of temperature and transmutation on any metal contained in thermocouple thermometers, ultimately providing the temperature drift behavior of the thermocouple. Experimentally, many researchers have conducted irradiation tests on common in-reactor thermocouples such as nickel-based, platinum-based, tungsten-based, and Mo–Nb thermocouples. These studies found that irradiation has a relatively small impact on the measurement accuracy and stability of nickel-based and Mo–Nb thermocouples, while the effects on platinum-based and tungsten-based thermocouples are significant.

2.3 Case studies of practical applications

Mo and Nb are the most promising thermocouple wire materials for resisting thermoelectromotive force drift (i.e., calibration drift) in high-temperature irradiation environments when placed in close proximity to or within nuclear fuel for long-term testing. For quantitative evaluation, HTIR-TCs were inserted into the Advanced Gas-cooled Reactor (AGR)-5/6/7 fuel test in a high-temperature gas-cooled reactor, undergoing in-pile testing for up to 12 months under high neutron flux and high-temperature conditions. The fuel test assembly consisted of five irradiation capsules placed in the northeast flux trap of the ATR core, which has an inner diameter of 13.34 cm. Each capsule has a diameter of approximately 7 cm (2.75 inches), and the five capsules were welded together to form a test string with a total length of 1.22 meters (the orientation of the test string and the layout of the thermocouples are shown in **Figure 3**).

The fuel capsules are arranged in bottom-to-top order: Capsule 1 is at the bottom of the reactor active zone, and Capsule 5 is at the top. Each capsule is loaded with different amounts of test fuel, which interacts with the neutron flux to produce different temperature gradients. The thermocouple (TC) system subsequently performs temperature measurements to obtain key test data on fuel performance and the performance of the thermocouples themselves. The thermal neutron flux in the ATR generally follows a symmetric cosine-squared distribution: at the midplane of the reactor height, the perturbed thermal neutron flux peaks at approximately 2.8×10^{14} n/cm$^2 \cdot$ s, while the fast neutron flux (E > 1 MeV) reaches about 2.25×10^{14} n/cm$^2 \cdot$ s. The neutron flux decreases in a cosine-squared pattern outward from the centerline and decays rapidly toward the top and bottom of the reactor. The temperature range of each fuel capsule depends on the total irradiation dose and its

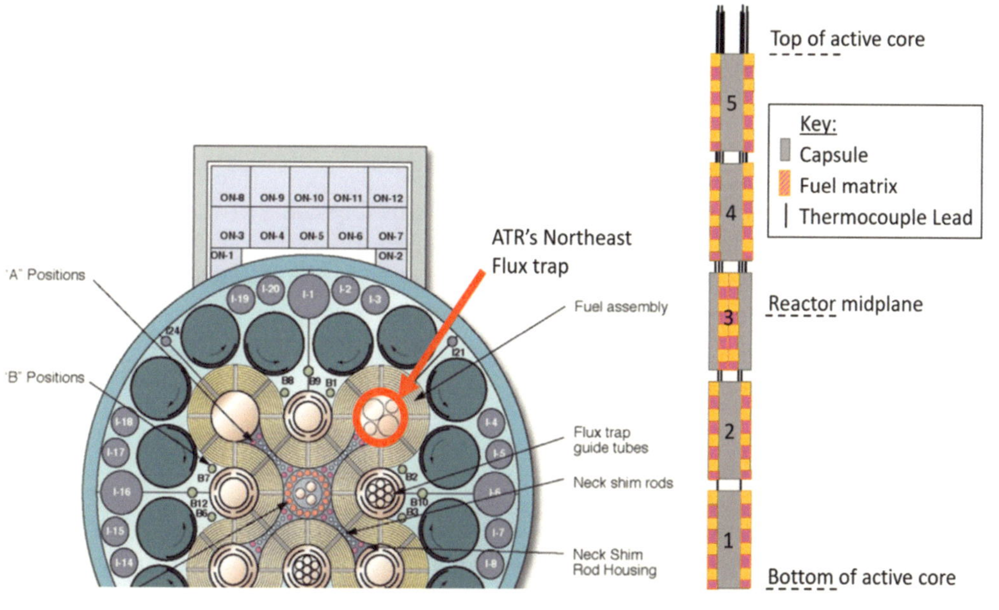

Figure 3.
Layout of temperature measurement capsules in an actual high-temperature gas-cooled reactor core.

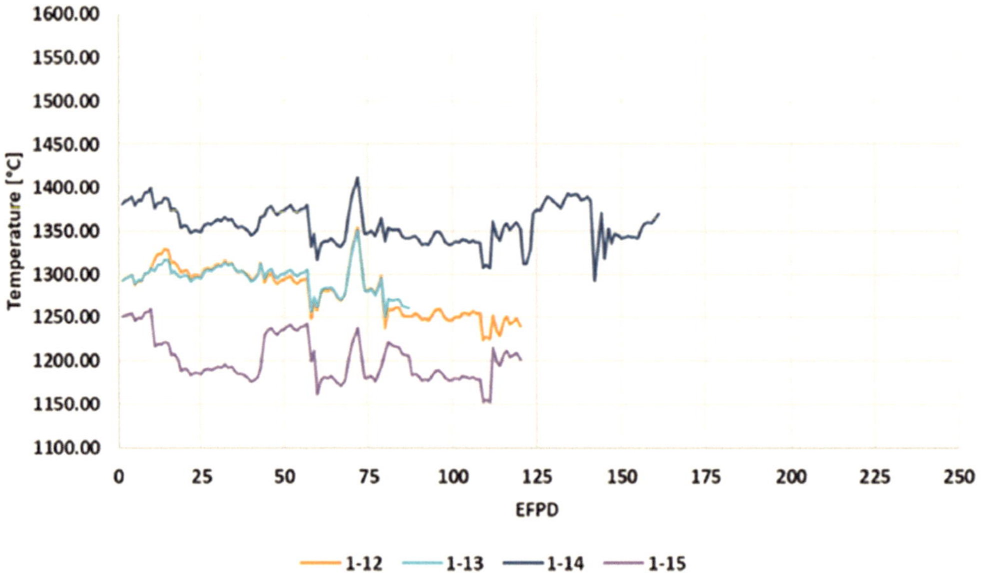

Figure 4.
Characteristics of temperature distribution.

position in the core, with Capsule 3, located at the midplane of the reactor height, expected to exhibit the highest temperature. Among the five fuel capsules, only Capsule 1 and Capsule 3 are equipped with HTIR-TCs for experimental temperature measurement. However, the lead conduits of each thermocouple must be routed upward through dedicated channels in the adjacent fuel capsules above to exit the core. This means that, for example, HTIR-TCs measuring the temperature of Capsule 1 must pass through high neutron flux regions of the core. Therefore, although the thermocouples in Capsule 1 operate at lower temperatures, the thermocouple drift caused by thermal and fast neutron flux is similar across all thermocouples in both Capsule 1 and Capsule 3. Using an array of innermost temperature sensors closest to the fuel, HTIR-TCs 1–12, 1–13, 1–14, and 1–15 measured the temperature in the hottest region of Capsule 1. Data acquisition lasted approximately 425 calendar days, covering the operational period of all HTIR-TCs in Capsule 1 of the AGR-5/6/7 test assembly. **Figure 4** shows the daily average temperature of each HTIR-TC during operation, calculated based on the effective operating time of each thermometer combined with the duration of ATR full-power operation[5].

3. Fiber optic temperature sensor

3.1 Measurement principle

The principle of optical fiber temperature measurement involves light waves from a source propagating through an optical fiber into a modulation region, where they are modulated by temperature, resulting in changes in optical parameters that describe the characteristics of the light wave – such as intensity, wavelength, frequency, phase, and polarization state – thereby converting it into

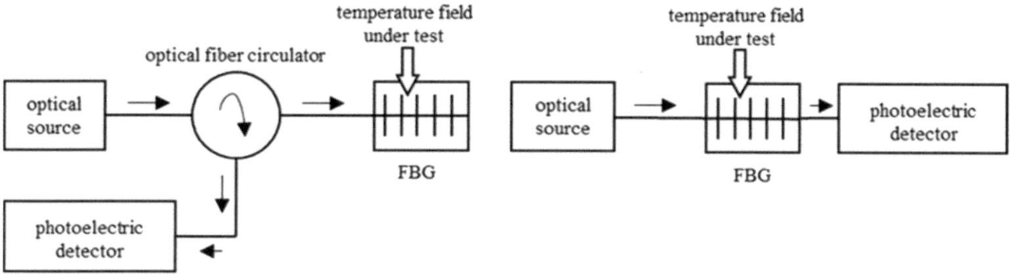

Figure 5.
Principles of temperature measurement using different FBGs.

modulated signal light. This signal is then transmitted via optical fiber to a photodetector, where a demodulator extracts the temperature-related information to ultimately determine the measured temperature. In the field of high-temperature measurement, optical fiber sensors offer advantages such as a broad temperature measurement range, fast thermal response, high accuracy, long service life, and excellent mechanical sensitivity. Furthermore, they are immune to strong electromagnetic interference and can operate in harsh environments including high temperature and high pressure. Based on their working principles, fiber optic temperature sensors are mainly divided into four categories: fiber grating sensors, which include Fiber Bragg Gratings (FBG), Long Period Fiber Gratings (LPFG), and Chirped Fiber Gratings (CFBG). These sensors rely on temperature-induced wavelength shifts or mode coupling for measurement and are suitable for applications such as transformer windings and aircraft engines, with femtosecond laser-inscribed FBGs capable of withstanding temperatures up to 1,200°C. FBGs are novel optical devices developed in the 1990s, primarily classified into short-period FBGs (Bragg FBGs) and long-period FBGs. Short-period FBGs act as reflective bandpass filters, while long-period FBGs function as transmissive bandstop filters. When an optical signal passes through an FBG, a spectral component with a central wavelength λ is separated via reflection or transmission. This wavelength λ is related to the refractive index of the medium, and changes in the external temperature cause corresponding shifts in λ. A spectrum analyzer determines the measured temperature by detecting these changes in λ [6]. The principle of FBG temperature measurement is illustrated in **Figure 5** [7, 8].

The majority of research on fiber-optic temperature measurement within reactors has focused on FBG technology. FBGs are among the most representative and widely used types of fiber gratings, also referred to as short-period gratings, with typical grating periods on the sub-micron scale. In FBGs, optical coupling occurs between modes propagating in opposite directions, making them reflective bandpass filters characterized by a narrow reflection band. The peak wavelength of this band is known as the Bragg wavelength. When light passes through a Bragg grating, it is strongly reflected at wavelengths that satisfy the phase-matching condition, while exhibiting weak reflection at wavelengths that do not meet this condition. The phase-matching condition is given by:

$$\lambda = 2n_0\Lambda \tag{1}$$

where n_0 represents the effective refractive index in the grating region, and Λ denotes the grating period. As can be seen from Equation (1), FBGs are intrinsically sensitive to temperature and strain by virtue of induced changes in n_0 and Λ, which makes them widely employed as temperature and strain sensors.

When the temperature changes, it induces variations in both the grating period and the effective refractive index, resulting in a shift in the Bragg wavelength. Therefore, the relationship between the wavelength shift $\Delta\lambda_T$ and the temperature change ΔT can be expressed as:

$$\Delta\lambda_T = \lambda \cdot \left(\frac{1}{\Lambda}\frac{\partial\Lambda}{\partial T} + \frac{1}{n_0}\frac{\partial n_0}{\partial T}\right) \cdot \Delta T \tag{2}$$

In the case of fiber gratings, temperature changes primarily induce the thermal expansion effect and the thermo-optic effect. The thermal expansion effect mainly influences the grating period, while the thermo-optic effect predominantly affects the effective refractive index. Thus, the relationship can be simplified as:

$$\Delta\lambda_T = \lambda \cdot (\alpha + \beta) \cdot \Delta T \tag{3}$$

Where α is the thermal expansion coefficient of the optical fiber material, and β is the thermo-optic coefficient, representing the rate of change of the refractive index with respect to temperature. As shown in Equation (3), within a certain temperature range where both α and β can be regarded as constants, the wavelength shift exhibits a proportional relationship with the change in temperature [9].

3.2 Research status and analysis

Berghmans *et al* [10]. compared three types of fiber-optic temperature sensors in a gamma radiation environment (160 kGy) in 1998. Semiconductor absorption sensors exhibited stable performance with errors below 1°C, though neutron radiation altered their absorption characteristics, causing deviations. Fabry–Pérot (F–P) cavity sensors showed monotonic decreases in readings due to radiation-induced wavelength-selective attenuation, while fluorescent sensors failed completely at 250 Gy due to radiation-induced attenuation, indicating that semiconductor absorption types are most suitable for gamma environments but require optimization for neutron radiation resistance. Gerrit J. de Villiers et al. [11] addressed the need for high-temperature distributed temperature monitoring in Pebble Bed Modular Reactor (PBMR) cores, theoretically and experimentally validating the feasibility of fiber Bragg grating (FBG) technology. Key results showed that Type I silica-based FBGs had a linear sensitivity of about 10 pm/°C below 300°C, but temperature gradients caused reflection spectrum broadening, affecting multiplexing accuracy; femtosecond laser-inscribed Type II FBGs showed no degradation at 1,000°C and exhibited a quadratic wavelength-temperature relationship, breaking the conventional FBG temperature limit; although sapphire FBGs withstand 2000°C and are radiation-resistant, their multimode transmission, high attenuation, and low reflectivity (only 5%) resulted in excessive signal noise, making them impractical for 30-meter long-distance applications. It was concluded that Type II silica-based FBGs are a practical solution for high-temperature nuclear reactor temperature measurement, while sapphire FBGs require signal quality improvements. In 2008, Sang et al. [12] used Rayleigh scattering for distributed temperature measurement in a nuclear reactor, achieving 1 cm spatial resolution and a maximum temperature of 850°C; copper-clad fibers had the highest temperature sensitivity, with higher doping concentrations increasing responsiveness, demonstrating that Rayleigh scattering technology is suitable for high-resolution in-reactor temperature measurement. In 2023, Hyer et al. [13] tested metal-armored distributed optical fibers in a simulated gas-cooled reactor environment, monitoring six gas channels (275–400°C) simultaneously with a single fiber at 1.6 mm spatial resolution, suppressing vibration via spot welding; radiation-induced drift required further optimization, but distributed sensing captured local temperature transients, improving core calorimetric accuracy. Fernandez et al. [14] tested FBG temperature sensors in SCK•CEN's BR1 (low flux) and BR2 (high flux) reactors in Belgium from 2001 to 2002. Results in the low-neutron-flux, air-cooled graphite BR1 reactor showed less than 1°C error compared to thermocouples for untreated FBGs. In the high-flux BR2 material test reactor, severe degradation occurred at 90°C with spectral broadening and decreased reflectivity, but low-temperature (50°C) FBGs remained stable under 160 MGy dose and 8×10^{18} n/cm^2 neutron fluence. Mixed gamma-neutron radiation induced higher sensitivity than gamma radiation alone, and hydrogen-loaded fiber FBGs showed greater radiation sensitivity. In 2004, Mihailov et al. [15] used an 800 nm femtosecond laser with a phase mask to inscribe FBGs in pure SiO_2 and germanium-doped fibers, achieving high refractive index modulation ($\Delta n \approx 1.9 \times 10^{-3}$) without hydrogen loading and stability at 950°C; the femtosecond laser-inscribed FBGs had low polarization-dependent loss (PDL < 0.24 dB), making them suitable for high-temperature nuclear environment sensors. Between 2016 and 2017, Morana et al. [16] developed a patented process for radiation-resistant FBGs (femtosecond laser inscription + 750°C thermal treatment) and validated it under X-ray/gamma radiation; under 1 MGy dose and 350°C, Bragg wavelength shift was below ± 10 pm (temperature error below ± 1°C). Fluorine-doped fibers performed even better with errors under ± 0.4°C, and stability was unaffected by dose rates (1–50 Gy/s); high reliability was also verified under neutron irradiation (5×10^{19} n/cm^2) and proton irradiation (63 MeV). In 2017, Kuhnhenn et al. [17] tested Morana team's radiation-resistant FBGs under gamma radiation (Co-60 source) from room temperature to 335°C; at 200 kGy dose, Bragg wavelength shift was within ± 2 pm, temperature error below ± 0.2°C, with stable performance at 335°C, confirming process reproducibility and suitability for real-time monitoring in nuclear facilities.

3.3 Case studies of practical applications

MYRRHA [18] is a prototype Generation IV nuclear reactor system. It is an accelerator-driven system cooled by molten lead-bismuth eutectic (LBE) within a temperature range from 125°C (the melting point of LBE) to 400°C. However, during operation, the core outlet temperature is expected to reach up to 700°C. As in any nuclear reactor, temperature must be monitored and controlled in every section or component to ensure safe operation, with the fuel assembly being one of the most critical components. In MYRRHA, a fuel assembly consists of a total of 127 parallel fuel rods equipped with wire spacers. **Figure 6** shows a subassembly comprising seven such fuel rods. These fuel rods are vertically arranged within the reactor. They are mechanically fixed and supported at the bottom but free at the top. The LBE coolant flows upward. A helical wire spacer is wound around each fuel rod to maintain adequate spacing during operation. These wire spacers restrict the gap

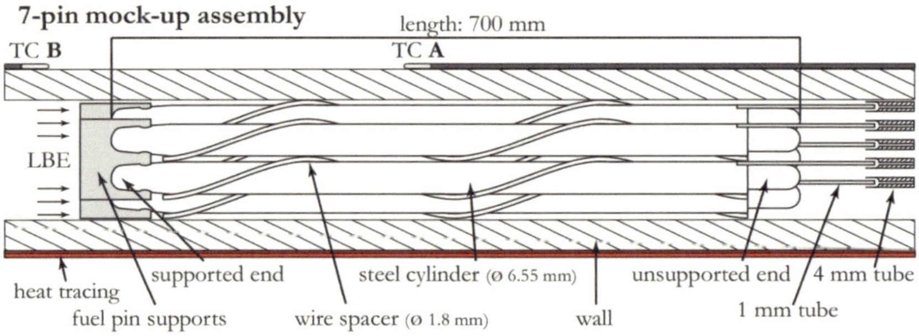

Figure 6.
Schematic diagram of the fuel assembly.

between individual fuel rods to just a few millimeters. To obtain reliable temperature readings inside the fuel assembly and use them for purposes such as reactor control, the sensors must not disturb the LBE flow. Therefore, any sensor introduced into the fuel assembly must be extremely small in size.

A LBE loop was constructed to expose a simulated fuel assembly to LBE flow under conditions representative of the MYRRHA reactor. The entire setup, shown in **Figure** 7, measures 3.5 m × 2.5 m and comprises an LBE storage tank, the test loop, and a control system. The loop holds approximately 350 kg of LBE and operates at temperatures up to 250°C. A 5.5 kW electric motor (model FCA 132SA-2) drives a magnetically coupled triple-screw pump (model Kral K 55–118) to achieve mass flow rates up to 20.5 kg/s. Heating and control are managed via a SCADA system integrating pressure sensors, tachometers, and 35 heating circuits – each with multiple trace heating lines – along with K-type thermocouples mounted on the loop housing, using a sampling interval of 30 seconds. The vertical test section downstream of the pump is a 0.767 m-long steel cylinder with an inner diameter of 51.2 mm, topped with a 0.450 m T-piece used to divert

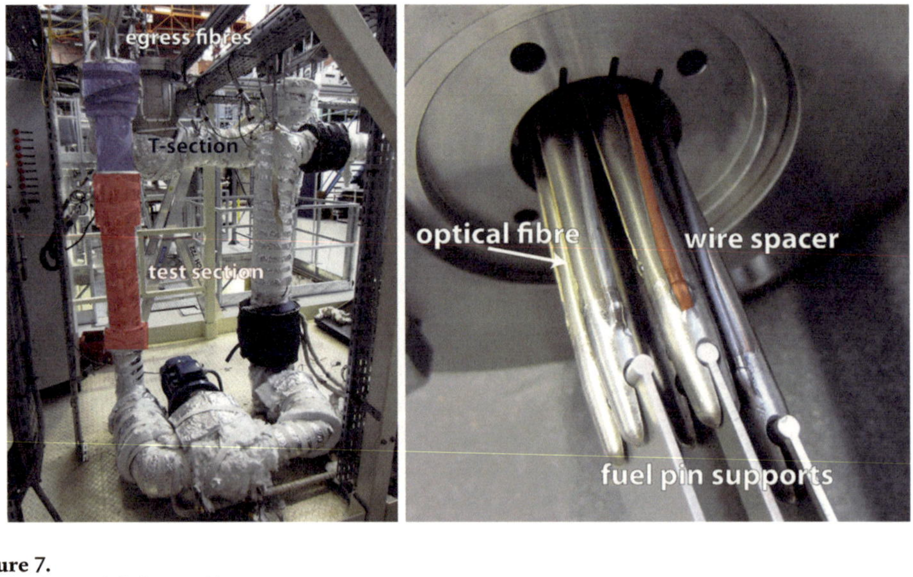

Figure 7.
Test facility and fuel assembly.

LBE flow and serve as the optical fiber exit. Fuel rod simulators are made of 316 L stainless steel tubes with a diameter of 6.55 mm and length of 700 mm, matching the diameter of actual MYRRHA fuel rods while at half the length. Each rod is helically wound with a 1.8 mm-diameter wire spacer at a pitch of 265 mm. The rod holders are fixed in the test section with flanged connections. To mimic the hexagonal arrangement of fuel rods within the closely surrounded bundle in MYRRHA, a hexagonal housing was placed around the fuel rod bundle.

Employing a spectroscopy-based demodulation approach offers particular advantages, such as enabling the use of state-of-the-art Bragg wavelength identification techniques. Since the expected wavelength fluctuations were small and peak shape distortions were anticipated, a phase-correlation-based method was selected to analyze the acquired spectra, allowing real-time recording and processing at acquisition rates of up to 5,000 spectra per second. Using the phase-correlation method, Bragg wavelength shifts as small as 1 pm could be resolved. As the FBGs were integrated inside the fuel rods, they were sensitive to both the thermal expansion of the rod itself and the effects of the adhesive and fiber coating. After the complete loop was properly sealed, all sections were heated and any residual oxygen was removed prior to filling the circuit with LBE. These procedures were continuously monitored using the FBG sensors. As the temperature in the test section increased, the fuel rods were heated, and the FBGs measured the corresponding thermal expansion. However, once the vacuum was established, the fuel rods became effectively thermally insulated, resulting in a more uniform temperature distribution along each rod. This effect is visible in **Figure 8**, which shows temperature measurements over time from multiple FBGs on a single fuel rod. It should be noted that since the FBGs only measure temperature changes, absolute temperatures must be derived using a temperature reference, such as thermocouple data. In this case, the reference was obtained from thermocouples mounted on the test section wall: once vacuum was achieved and temperature distribution became more uniform, the test section interior and wall were assumed to be nearly isothermal. Therefore, the uncertainty in the absolute temperature measurement using this reference is determined by the measurement uncertainty of the K-type thermocouples,

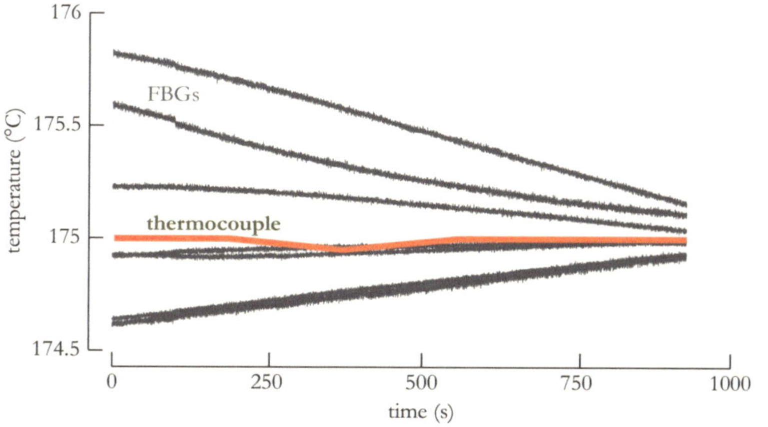

Figure 8.
Temperature measurement of a single fuel pellet.

which is approximately ± 2 K. The FBG data allowed qualitative verification of the vacuum integrity inside the loop – an effect not discernible from the thermocouple readings, as they were placed externally on the test section and thus more influenced by the ambient environment.

The FBG sensors represent the first successful application of temperature sensing within an LBE-cooled nuclear fuel assembly, overcoming challenges related to extreme space constraints, high corrosion, and electromagnetic interference. With a resolution of 30 mK, they significantly outperform conventional thermocouples, which typically have an accuracy of around ± 2 °C. Furthermore, the system supports multi-parameter monitoring capabilities including temperature distribution, liquid level variation, and equipment status (e.g., vacuum integrity and fault detection). This technology is directly applicable to temperature control and safety systems in liquid metal-cooled reactors such as MYRRHA, and can be extended to monitor other high-temperature and high-radiation components.

4. Passive thermoacoustic sensor

4.1 Measurement principle

The Passive Thermoacoustic Sensor utilizes the thermoacoustic effect to convert temperature signals into acoustic waves, enabling passive temperature measurement. The thermoacoustic effect describes the mutual conversion between thermal energy and acoustic energy. Acoustic vibrations amplify when heat is supplied during the densest phase and removed during the sparsest phase; conversely, vibrations decay if heat is removed at densest phase and supplied at sparsest phase. This device applies the amplification principle from Rayleigh's criterion [19].

The fundamental principle of the passive temperature measurement device in a PWR core is illustrated in **Figure 9**. During operation, the hot end absorbs heat from fuel rods, creating a temperature gradient across the stack. The cold end maintains ambient temperature. When gas moves leftward (Step ①), heat transfers to the gas; when gas moves rightward (Step ②), heat releases back. This cyclic heat exchange generates high-frequency gas vibrations, producing intense acoustic waves in the resonator. In Step ②, heat transfer from the stack to the gas increases gas temperature ($T^{++} \rightarrow T^{+++}$) and pressure. The resulting expansion under high pressure performs work ($p\Delta V$). Pressure rise pushes gas further during each cycle. During rightward movement (Step ③), heat transfers back to the stack (Step ④), reducing gas temperature ($T^{+} \rightarrow T_0$) and pressure. This heat removal under low pressure contracts the gas, performing additional work ($p\Delta V$). Ultimately, acoustic wave amplitude stabilizes when energy dissipation balances thermoacoustic conversion, sustaining an acoustic pressure wave [20].

Figure 10 shows a nuclear fuel rod with a stack structure for thermoacoustic sensor operation. Gas oscillating in the resonator tube has frequency f determined by tube length L and gas sound speed c. For a uniform tube in fundamental half-wavelength mode, $f = c/2l$. Since sound speed relates to absolute temperature T (in K) by $c = \sqrt{(kRT)}$ (where k is adiabatic index, R is gas constant), the standing wave frequency correlates with temperature. This "thermoacoustic thermometer" converts reactor heat into acoustic oscillations without electrical power.

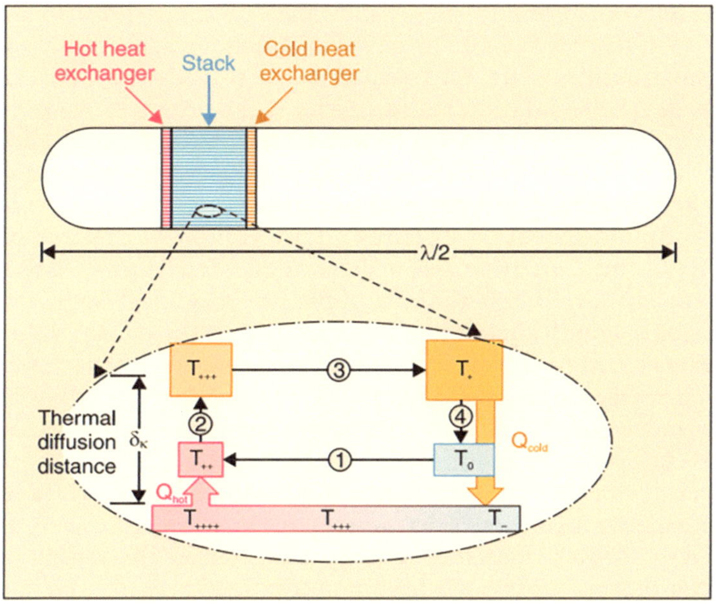

Figure 9.
Temperature measurement principle diagram.

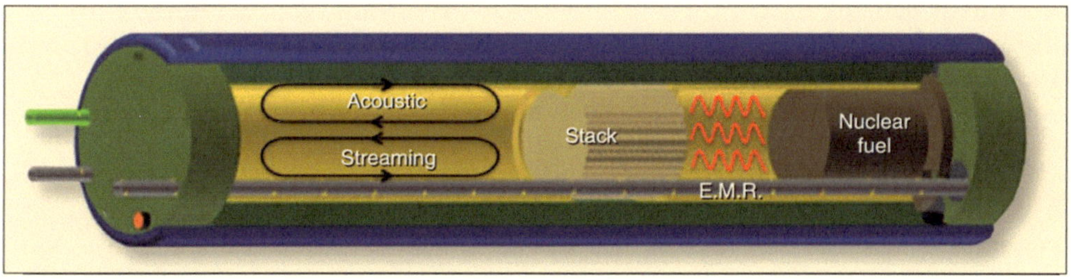

Figure 10.
Schematic diagram of a fuel rod including a stack.

Temperature data transmits remotely via coolant, making it ideal for passive-safety-critical nuclear applications [19].

4.2 Research status and analysis

Passive instrumentation fundamentally eliminates this vulnerability by operating without power supplies or complex circuits. Research into passive instruments is thus essential for enhancing the reliability of passive safety systems and overall nuclear plant security. The development of such power-independent monitoring devices will significantly strengthen the robustness of instrumentation systems [21]. Traditional thermocouples face multiple degradation mechanisms in high-ionizing-radiation environments such as nuclear reactor cores and spent fuel storage pools: Neutron bombardment induces lattice distortion in thermoelectric materials, causing Seebeck coefficient drift.

For instance, K-type thermocouples exhibit measurement deviations exceeding 5°C at fast neutron fluences of 10^{18} n/cm^2. Concurrently, gamma-ray ionization degrades insulation materials. Magnesium oxide (MgO) insulation layers suffer a precipitous drop in insulation resistance by three orders of magnitude at doses of 10^6 Gy, creating significant signal short-circuit risks. Structurally, hydrogen generated from radiolytic decomposition of water permeates metal grain boundaries, accelerating hydrogen embrittlement and cracking in sheaths. For example, 304 stainless steel experiences a 1.5% volume expansion at neutron fluences of 10^{23} n/m^2. Radiation also accelerates corrosion processes and degrades mechanical strength, reducing thermocouple lifespan under high-temperature and high-pressure conditions to merely one-third of the design value. These combined effects lead to reduced temperature measurement accuracy, response delays, and increased unplanned reactor shutdown risks. The Fukushima accident serves as a prime example, where radiation-induced damage critically compromised temperature monitoring capabilities.

In contrast, thermoacoustic sensors utilize materials inherently immune to radiation damage while uniquely harnessing high-energy neutrons and gamma rays as an energy source. Radiation energy absorbed by the sensor materials is converted into thermal energy. Subsequent localized heating and cooling cycles within the material induce periodic thermal expansion and contraction. This periodic mechanical excitation generates detectable acoustic waves (pressure waves) in the surrounding gas medium. Crucially, the frequency of these acoustic waves correlates directly with the material's temperature. External microphones or acoustic sensors detect these audio signals, enabling remote temperature inference within the material environment. The fundamental theory relies on the thermodynamic relationship between gas sound speed and temperature, expressed through the acoustic virial equation for gas within the resonant cavity:

$$u^2 = \frac{\gamma RT}{M}\left(1 + \frac{\beta_a}{RT}p + \frac{\gamma_a}{RT}p^2 + \cdots\right) \tag{4}$$

where u represent sound speed, p represent pressure, R represent gas constant, T represent temperature, M represent molecular weight, γ represent specific heat ratio, β_a and γ_a represent acoustic virial coefficients.

Within a cylindrical resonator cavity, sound waves propagate as longitudinal waves. When the frequency of the incident sound wave precisely matches the natural frequency of the gaseous medium inside the cavity, the gas undergoes resonance. Under this condition, the sound wave manifests as a standing wave pattern, and its amplitude undergoes significant amplification. According to fundamental acoustic theory, the relationship between the sound speed u and the ideal resonance frequency f_{lmn}^0 is expressed as:

$$u^2 = \frac{\gamma RT}{M}\left(1 + \frac{\beta_a}{RT}p + \frac{\gamma_a}{RT}p^2 + \cdots\right) u = \frac{2\pi f_{lmn}^0}{\sqrt{\left(\frac{l\pi}{L}\right)^2 + \left(\frac{x_{mn}}{a}\right)^2}} \tag{5}$$

Where a represent internal radius of cylindrical cavity; L represent internal length; l, ∣m∣, n = 0,1,2,. . .eigenvalues representing axial, azimuthal, and radial half-wave integers; x_{mn} represent the n–th root of $d_{J_m(x)}/d_x=0$; J_m represent the m–th order cylindrical Bessel function. The fundamental axial acoustic mode (100), characterized by the minimal frequency separation interval, is employed for measurement:

$$u = 2Lf_{100}^{0} \tag{6}$$

The experimentally measured acoustic resonance frequency f_N deviates from the ideal resonance peak f_{100}^{0} under the influence of non-ideal perturbations, primarily [21].

4.3 Practical application cases

Acoustic-thermal sensors possess two significant advantages: passive safety and radiation resistance, and hold great application prospects in multiple fields.

Among these, the thermoacoustic power sensor (TAPS) prototype used for performance evaluation in a sodium environment is shown in **Figure 11**. This prototype mainly consists of the following core components: slotted tube, TAPS Enclosure, electric heater, stack, Insulation Cap, Isolation Springs, Fixing Assembly, Thermocouple. TAPS Enclosure is an acoustic resonant cavity made of SS-316 L stainless steel, which complies with chemical compatibility requirements. It has undergone a leak-tightness test and is filled with pressurized inert gas to reduce oxidation of the electric heater and extend its service life. A section of stainless steel tube is welded to the hot end of the enclosure; this tube passes through the slotted tube and is fixed to it by welding. The wires of the electric heater are introduced through this tube. Electric heater is used to simulate the heating effect generated by fuel or gamma radiation energy harvesters in commercial TAPS. The acoustic stack is composed of ceramic blocks with tiny, continuous, and longitudinally arranged channels inside, through which gas mixtures (sound waves) can flow. A temperature gradient is established between the hot end and the cold end to initiate the thermoacoustic conversion process that converts thermal energy into acoustic energy. Springs are installed at the top and bottom inside the slotted tube, between the enclosure and the tube. They allow the TAPS enclosure to move with the thermoacoustic vibrations generated internally and transmit sound waves through the cooling medium (e.g., sodium) surrounding the TAPS. Thermocouple: A Type K thermocouple is inserted into the hot end of the TAPS prototype to monitor the temperature of the electric heater [22].

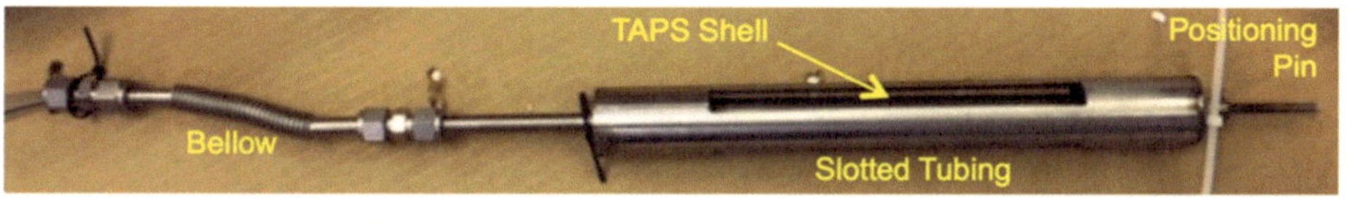

Figure 11.
The TAPS prototype for performance evaluation in a sodium environment.

5. Thermochromic liquid crystal thermometry

5.1 Measurement principles

Liquid crystals represent a state of matter intermediate between solids and liquids, exhibiting characteristics of both crystals and liquids. Their appearance resembles a fluid, demonstrating mechanical properties such as flowability, continuity, and surface tension, while their molecular structure is analogous to crystals, displaying physical properties like optical anisotropy and interference typical of solids. Thermochromic liquid crystal (TLC) thermometry is a non-contact temperature measurement technique. As an optically active chiral nematic phase material, as shown in **Figure 12**, the molecules of TLCs are arranged in layered structures, with each layer of molecules undergoing helical rotation along the principal axis. The pitch of the helix (i.e., the distance between two adjacent layers of liquid crystal molecules) determines the wavelength of reflected light. When changes in ambient temperature alter the interlayer molecular spacing of the TLCs, the frequency of the light waves selectively reflected by the liquid crystals also changes, resulting in a corresponding shift in the color displayed on the surface. The unique property of selective light reflection characteristic of chiral nematic phase materials forms the basis of TLC thermometry. Each displayed color corresponds to a specific temperature, allowing the temperature to be determined based on the observed color of the liquid crystals.

TLCs display distinct colors at different temperatures, characterized by a transition from colorless to red upon reaching the lower threshold of the active temperature range. As the temperature increases, the colors shift sequentially through orange, yellow, green, blue, and violet. Upon reaching the upper limit of the temperature range, the liquid crystals revert to a colorless state. Each color exhibited by the TLCs corresponds to a specific temperature, and a color-temperature correlation curve can be established through calibration experiments [23]. Based on the temperature range over which color changes occur, TLCs can be categorized into broad-band and narrow-band

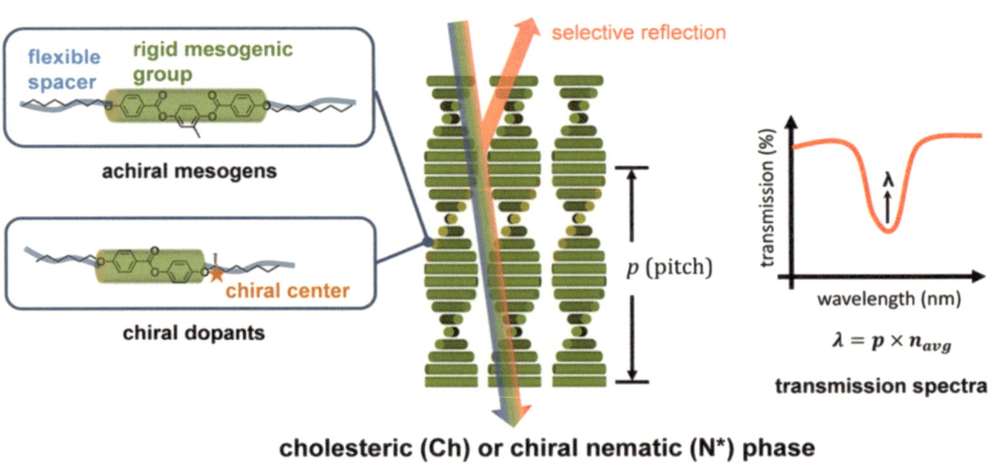

Figure 12.
Schematic diagram of the structure of chiral nematic liquid crystals.

types. Broad-band liquid crystals typically operate over an effective temperature range of 5°C to 30°C, offering advantages such as a wide measurement range, straightforward calibration, and suitability for analyzing dynamic temperature variations. In contrast, narrow-band liquid crystals function within a narrower range of 1°C to 2°C, providing higher measurement accuracy, easier data fitting, and reduced sensitivity to external lighting conditions and illumination angles [24].

5.2 Research status and analysis

TLC thermometry is a non-contact measurement technique characterized by high resolution (temperature resolution up to 0.1°C, spatial resolution up to 1 mm), rapid response, and reversible color changes of the TLCs. Consequently, numerous domestic and international scholars have conducted extensive research on the calibration and heat transfer applications of TLCs.

Since traditional thermocouple contact measurement methods may disturb flow fields and introduce measurement errors, Chyu [25] introduced a method for obtaining local heat transfer coefficients using transient liquid crystal thermography. This study used the inlet temperature as a reference and evaluated four methods for determining the local overall average temperature in turbine blade cooling. The research highlighted the potential of liquid crystals for detailed thermal measurements in complex geometries. Bunker's team [26] utilized TLC thermometry to conduct a detailed study of the heat transfer coefficient and pressure loss in a 45° staggered ribbed cooling channel within a temperature range of 40–45°C and a Reynolds number (Re) range of 20,000–100,000. The study accurately captured temperature variations in the cooling channel and derived corresponding changes in the heat transfer coefficient. Taslim's team [27] employed TLC technology to investigate the heat transfer coefficient in a spanwise rotating channel with rib-roughened walls, as well as the effects of rotation, aspect ratio, and rib roughness on heat transfer performance in rib-roughened channels. The study accurately obtained the heat transfer characteristics of ribbed channels under rotating conditions, demonstrating that liquid crystal thermometry is an effective and accurate method for determining heat transfer coefficients in non-stationary test sections. Kumar's team [28] developed an innovative method using TLCs and Bayesian inference to detect hotspots and estimate the strength of heat sources in a flat-plate system. The experimental setup involved discrete strip heat sources embedded in a flat plate, with temperature distributions measured via TLC sheets. Numerical validation of the experimental data showed discrepancies within 3%, confirming the robustness of the method. Schmid's team [29] proposed an advanced transient heated foil method using TLCs to measure local heat transfer coefficients. By employing a linearly increasing surface heat flux, the method improved accuracy and addressed challenges in low heat transfer regions. Validation was performed on a flat plate subjected to a single circular jet impingement, with results showing good agreement with established correlations in the literature. The study emphasized the adaptability of this method for applications requiring high-resolution thermal diagnostics, provided that appropriate optical access and surface geometry are maintained.

5.3 Practical application cases

Phase change materials (PCMs) play a critical role in latent heat thermal energy storage (LHTES) systems, and the optimization of their thermal performance is central to improving energy conversion efficiency. While the addition of nanofillers enhances thermal conductivity, it also alters the material's optical properties, rendering traditional phase interface visualization methods ineffective. Therefore, developing novel temperature measurement techniques suitable for opaque nano-enhanced phase change material (NePCM) to accurately capture temperature distributions during phase change processes holds significant theoretical and engineering value for deepening the understanding of heat transfer mechanisms and optimizing energy storage performance. Current research primarily relies on thermocouples to measure heat flux and fluid flow distribution details at specific moments, but this method suffers from low spatial resolution, and its invasive nature can introduce various artifacts that affect experimental results. Li [30] employed TLC technology to measure the temperature of invisible phase interfaces, investigating the melting process of NePCM in a differentially heated rectangular cavity. This approach enabled direct observation of the transition in heat transfer mechanisms from conduction to convection during melting and accurately determined instantaneous melt fraction and total heat storage.

Figures 13(a), **(b)**, and **(c)** present schematic diagrams of the external and internal structural components of the experimental setup, including a constant-temperature bath, a rectangular cavity test section, a camera with an integrated illumination source, two aluminum plates, two constant-temperature water circulators, and a data acquisition unit. **Figure 13(b)** details the rectangular cavity test section, which consists of multiple poly plates and two copper blocks (serving as heating and cooling boundaries) connected via flanges.

The heating and cooling boundary temperatures were controlled by aluminum plates with internal flow channels, connected to water baths with a temperature stability of 0.01°C. Thermal grease was applied between the copper blocks and aluminum plates to reduce interfacial thermal resistance. As shown in Figure 22(c), except for the front optical observation window, the test cavity was wrapped with a thick insulating sleeve to minimize heat loss. T-type thermocouples with a calibration accuracy of ± 0.2°C were inserted into the copper plates to monitor the heating and cooling boundary temperatures, while two additional thermocouples were inserted into the top insulation layer to roughly estimate heat loss from the adiabatic walls.

From the temperature variation maps derived using the TLC thermography method, which reflect the thermal response of the samples during melting, the dynamic evolution and movement of the phase interface for different NePCM samples can be observed. As shown in **Figure 14**, in cavities with aspect ratios of H/B = 0.8, the melting fronts of three NePCM samples with different nanoparticle loadings (0 wt.%, 1 wt.%, and 3 wt.%) exhibit significant differences.

During the initial stage of melting in NePCMs, the melting fronts of all samples exhibited a vertical morphology, indicating dominant heat conduction. Notably, the 3 wt.% high-loading sample demonstrated the fastest advancement of the melting front due to its enhanced effective thermal conductivity (101% improvement compared to pure PCM). As melting progressed, natural convection effects gradually emerged in the molten region, characterized by distinctive bending of the upper portion of the melting front (as observed in the 20-minute image). This deformation

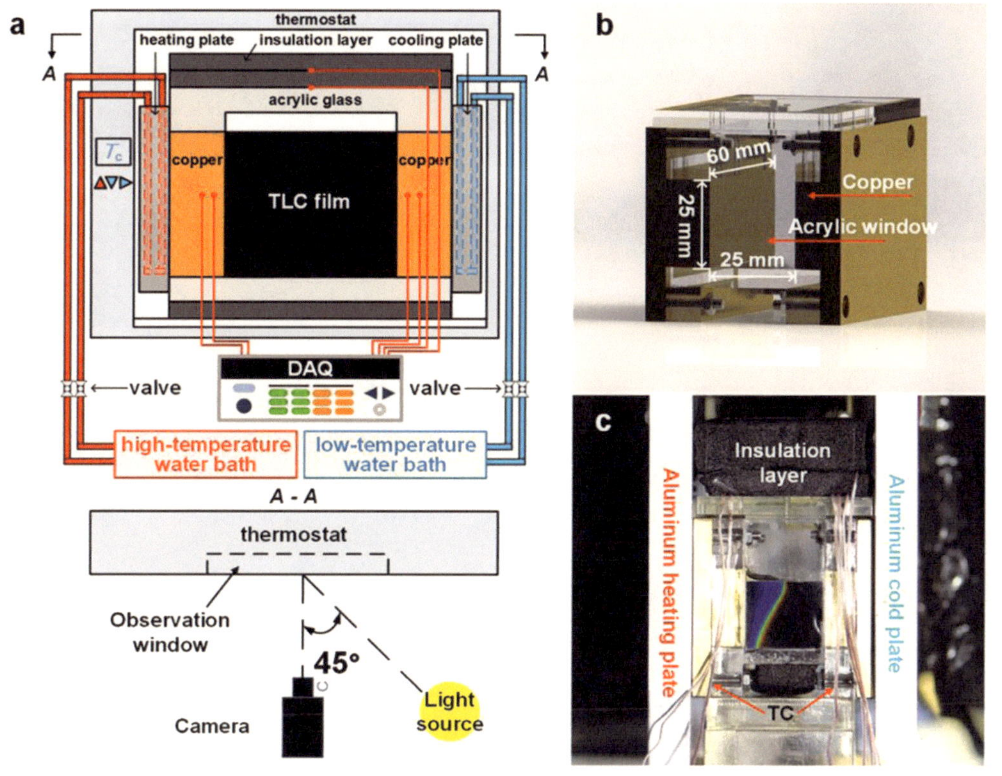

Figure 13.
Schematic diagrams of experimental components: (a) schematic of the experimental setup, (b) three-dimensional model of the test section (i.e., The differentially heated rectangular cavity), (c) photograph showing the assembly of the rectangular cavity.

originated from the buoyant rise of thermal plumes near the heated wall, forming a clockwise vortex at the top of the cavity.

6. Conclusions

This section introduces four types of advanced temperature measurement sensors. Among them, sheathed thermocouple temperature sensors represent the most mature technology. Mo and Nb thermocouples can be selected for irradiation environments, enabling stable and accurate temperature measurement in high-radiation conditions. However, as a contact-based method, it may disturb the internal temperature field and imposes certain requirements on measurement positioning. Fiber-optic temperature sensors, known for their flexibility and fast response, are suitable for monitoring in confined spaces and highly corrosive environments. FBGs exhibit the broadest applicability, though radiation environments significantly impact their performance. Radiation resistance can be enhanced through germanium doping. TLC thermometry is a non-contact method that utilizes the effect of temperature on liquid crystals to produce different colors corresponding to varying light wavelengths, thereby indicating temperature

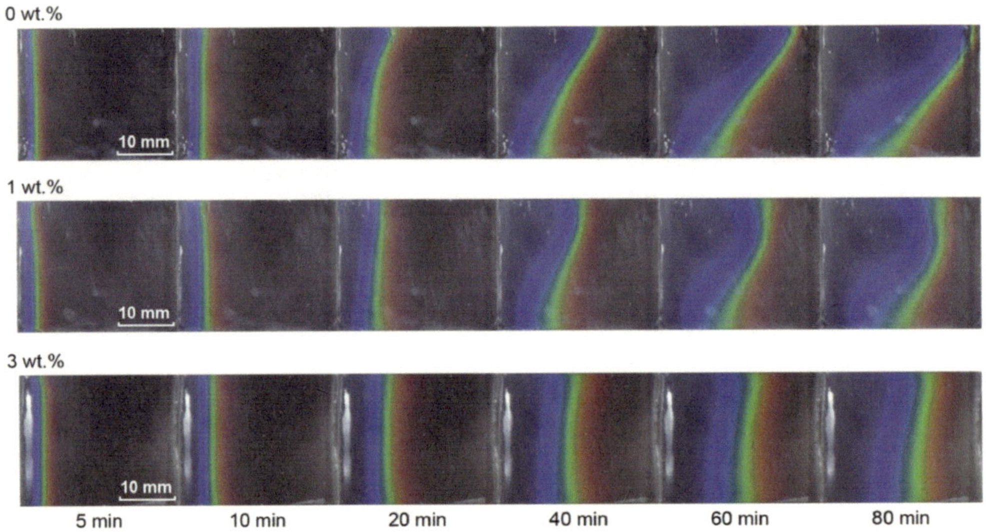

Figure 14.
Evolution of the phase interface during melting of NePCM samples with different nanoparticle loadings at an aspect ratio of H/B = 0.8.

changes. However, this method has limited radiation resistance and is primarily applied in radiation-free environments within nuclear reactors. Acoustic thermometry sensors leverage the thermoacoustic effect to convert temperature signals into acoustic signals, enabling self-powered temperature measurement. This approach can utilize ionization radiation energy by absorbing and converting it into heat, achieving radiation-resistant temperature measurement. Currently, this method is applied in liquid metal fast reactors.

Finally, a comprehensive comparison of the four temperature measurement methods is presented in **Table 1**. in the field of advanced temperature measurement, sheathed thermocouples and fiber Bragg grating sensors have reached a considerable level of maturity and are ready for practical application. Among them, sheathed thermocouples, particularly those made of Mo and Nb, have become a mainstay for temperature measurement in high-radiation environments due to their stability and reliability. Meanwhile, fiber optic sensors have gained widespread use

Sensor Type	Measurement Range	Accuracy	Radiation Resistance	Response Time
Sheathed Thermocouple	Very Wide	High, Stable	Strong	Medium
Fiber Optic Sensor	Wide	High	Strong	Fast
Thermochromic Liquid Crystal	Narrow	Medium	Weak	Fast
Acoustic Thermometry	Wide	Medium	Very Strong	Slow

Table 1.
Comparison of four types of temperature sensors.

in non-radiative, confined spaces and corrosive environments, leveraging their flexibility, corrosion resistance, and fast response. However, potential and challenges coexist: acoustic temperature sensors demonstrate revolutionary prospects. Utilizing the thermoacoustic effect to enable self-powered and non-contact measurement, they can even harness radiation energy, showing great potential for extreme environments where other sensors struggle to survive. Nevertheless, further research is still needed to improve their system accuracy, response speed, and engineering simplification. Similarly, TLCs are irreplaceable for providing high-resolution full-field temperature visualization, but their fatal flaw of extremely weak radiation resistance severely limits their application in nuclear environments, urgently requiring breakthroughs in materials science. In summary, the core gaps that remain are as follows: we still lack ideal sensors that can simultaneously deliver high accuracy, fast response, and exceptional environmental robustness. At the same time, significant technological voids still need to be filled in achieving panoramic 3D temperature field visualization in extreme environment sand enabling self-powered and maintenance-free operation throughout their entire lifecycle.

Acronyms and abbreviations

AGR	Advanced Gas-cooled Reactor
AP1000	Advanced Passive PWR 1000
ATR	Advanced Test Reactor
CAP1400	China Advanced Passive PWR 1400
CFBG	Chirped Fiber Gratings
FBG	Fiber Bragg Gratings
HTIR-TC	High-temperature irradiation-resistant thermocouple
LBE	Lead-bismuth eutectic
LHTES	Heat thermal energy storage
LPFG	Long Period Fiber Gratings
LMFBR	Liquid Metal Fast Breeder Reactor
Mo	Molybdenum
Nb	Niobium
NePCM	nano-enhanced phase change material
PBMR	Pebble Bed Modular Reactor
PCMs	Phase change materials
PWR	Pressurized Water Reactor
Re	Reynolds number
SCK•CEN	Studiecentrum voor Kernenergie/Centre d'Étude de l'énergie Nucléaire
TAPS	Thermoacoustic power sensor
TC	Thermocouple
TLC	Thermochromic liquid crystal

Author details

Haicai Lyu[1*], Xiaoyang Lun[1], Xianjun Chen[1], Pengcheng Yang[1], Mingqiang Yi[2] and Fenglei Niu[2]

1 School of Nuclear Science and Engineering, North China Electric Power University, Beijing, China

2 Microfluidic Foundry, San Pablo, California, USA

*Address all correspondence to: hc.lyu@ncepu.edu.cn

References

[1] Palmer AJ, Skifton RS, Scervini M, et al. Performance of custom-made very high temperature thermocouples in the advanced gas reactor experiment AGR-5/6/7 during irradiation in the advanced test reactor[C]. In: EPJ Web of Conferences. EDP Sciences. 2020; Vol. **225**. p. 04010

[2] Ross CW.Effect of thermal neutron irradiation on thermocouples and resistance thermometers[J]. Transactions of the American Institute of Electrical Engineers, Part I: Communication and Electronics. 1962;**81**(3):192–196

[3] Scervini M, Rae C, Lindley B. Transmutation of Thermocouples in Thermal and Fast Nuclear reactors[C]. In: 3rd International Conference on Advancements in Nuclear Instrumentation, Measurement Methods and Their Applications (ANIMMA). Marseille, France; 2013

[4] Skifton RS, Brian J, Scott R. High temperature irradiation resistant thermocouple drift model application to commercial thermocouples[R]. Idaho Falls: Idaho National Lab (INL); 2022

[5] Skifton R. High-temperature irradiation-resistant thermocouple instability model for in-pile reactor use[J]. Frontiers in Energy Research. 2023;**11**:1099584

[6] Ouyang K, Si M, Xing L, et al. Research progress and prospect of thermometry applied for in-pile temperature measurement in nuclear reactors[J]. China Measurement & Test 2025;**51**:1–18

[7] Tong X, Song L.Research status and application prospect of high-temperature optical fiber sensors [J]. Optical Communication Technology. 2014;**38**(10):8–10

[8] Han G, Liu X, Lei X, et al. Application of optical fiber sensing in aero-engine temperature test [J]. Chinese Journal of Scientific Instrument. 2022;**43**:145–164

[9] Liao G, He J, Zhu Y, et al. Fiber Grating Sensor and Application [J]. Equipment Environmental Engineering. 2022;**19**:142–149

[10] Berghmans F, Fernandez AF, Brichard B, et al. Radiation hardness of fiber-optic sensors formonitoring and remote handling applications in nuclear environments[J]. Proceedings of SPIE - the International Society for Optical Engineering. 1998;**3538**:28–39

[11] De Villiers GJ, Treurnicht J, Dobson RT. In-core high temperature measurement using fiber-Bragg gratings for nuclear reactors[J]. Applied Thermal Engineering. 2012;**38**:143–150

[12] Sang AK, Froggatt ME, Gifford DK, et al. One centimeter spatial resolution temperature measurements in a nuclear reactor using Rayleigh scatter in optical fiber[J]. IEEE Sensors Journal 2008;**8**:1375–1380

[13] Holden CH, Dominic RG, Christian MP. Toward local core outlet temperature monitoring in gas-cooled nuclear reactors using distributed fiber-optic temperature sensors[J]. Applied Thermal Engineering. 2023;**230**:120847

[14] Fernandez AF, Gusarov A, Brichard B, et al. Temperature

monitoring of nuclear reactor cores with multiplexed fiber Bragg grating sensors[J]. Optical Engineering. 2002;**41**:1246–1254

[15] Remy L, Cheymol G, Gusarov A, et al. Compaction in optical fibres and fibre Bragg gratings under nuclear reactor high neutron and gamma fluence[J]. IEEE Transactions on Nuclear Science. 2016;**63**: 2317–2322

[16] Morana A, Girard S, Marin E, et al. Radiation hardened fiber Bragg grating based sensors for harsh environments[J]. IEEE Transactions on Nuclear Science. 2017;**64**:68–73

[17] Kuhnhenn J, Weinand U, Morana A, et al. Gamma radiation tests of radiation-hardened fiber Bragg grating based sensors for radiation environments[J]. IEEE Transactions on Nuclear Science. 2017;**64**:2307–2311

[18] De Pauw B, Lamberti A, Ertveldt J, et al. Temperature monitoring using fibre optic sensors in a lead bismuth eutectic cooled nuclear fuel assembly[J]. Nuclear Engineering and Design. 2016;**297**:54–59

[19] Liu H, Xu J, Li Y, et al. A new passive thermoacoustic temperature measuring device for reactor core of pressurized water reactor [J]. Nuclear Safety. 2015;**14**:24–29

[20] Ali RA, Garrett SL, Smith JA, et al. Thermoacoustic thermometry for nuclear reactor monitoring[J]. IEEE Instrumentation & Measurement Magazine. 2013;**16**:18–25

[21] Li M, Xing L, Si M, et al. the investigation of a miniaturized acoustic gas thermometer [J]. Acta Metrologica Sinica. 2024;**45**:150–156

[22] Chien HT, Sienicki JJ, Elmer TW, et al. Demonstration of Thermoacoustic Power Sensor[R]. Argonne, IL (United States): Argonne National Lab.(ANL); 2019

[23] Lui X, Feng J. A new method of temperature measurement themochromic liquid crystal method [J]. Journal of Guangdong University of Technology. 2006;(04):50–53

[24] Zhang W. Design of Transient Thermal Liquid Crystal Calibration System and Multi Factor Error Analysis of Temperature Measurement [D]. Harbin(China): Harbin Engineering University; 2017

[25] Chyu MK, Ding H, Downs JP, Soechting FO. Determination of local heat transfer coeÅcient based on bulk mean temperature using a transient liquid crystals technique. n.d

[26] Bunker RS. Latticework (Vortex) cooling effectiveness: Part 1 — stationary channel experiments. In: Vol. 3 Turbo Expo 2004. Vienna, Austria: ASMEDC; 2004, p. 909–918

[27] Taslim ME, Bondi LA, Kercher DM. An experimental investigation of heat transfer in an orthogonally rotating channel roughened with 45 deg criss-cross ribs on two opposite walls. Journal of Turbomachinery. 1991;**113**:346–353

[28] Kumar S, Jakkareddy PS, Balaji C. A novel method to detect hot spots and estimate strengths of discrete heat sources using liquid crystal thermography. International Journal of Thermal Sciences. 2020;**154**:106377

[29] Schmid J, Gaffuri M, Terzis A, Ott P, Wolfersdorf JV. Transient liquid crystal thermography using a time

varying surface heat flux. International Journal of Heat and Mass Transfer. 2021;**179**:121718

[30] Li Z-R, Hu N, Liu J, Zhang R-H, Fan L-W. Revisiting melting heat transfer of nano-enhanced phase change materials (NePCM) in differentially-heated rectangular cavities using thermochromic liquid crystal (TLC) thermography. International Journal of Heat and Mass Transfer. 2020;**159**:120119